D1766867

A Passage to Infinity

A Passage to Infinity

Medieval Indian Mathematics from Kerala and Its Impact

GEORGE GHEVERGHESE JOSEPH

First published in 2009 by

SAGE Publications India Pvt Ltd
B1/I-1 Mohan Cooperative Industrial Area
Mathura Road, New Delhi 110 044, India
www.sagepub.in

SAGE Publications Inc
2455 Teller Road
Thousand Oaks
California 91320, USA

SAGE Publications Ltd
1 Oliver's Yard
55 City Road
London EC1Y 1SP, United Kingdom

SAGE Publications Asia-Pacific Pte Ltd
33 Pekin Street
#02-01 Far East Square
Singapore 048763

Published by Vivek Mehra for SAGE Publications India Pvt Ltd, Phototypeset in 10.5/12.5 AGaramond by Tantla Composition Services Private Limited, Chandigarh and printed at Chaman Enterprises, New Delhi.

Library of Congress Cataloging-in-Publication Data
Joseph, George Gheverghese.
A passage to infinity: medieval Indian mathematics from Kerala and its impact/George Gheverghese Joseph.
p. cm.
Includes bibliographical references and index.
1. Mathematics—India—Kerala—History. 2. Mathematics, Medieval.
3. Astronomy, Medieval—India—Kerala. I. Title.
QA27.I4J67 510.954′830902—dc22 2009 2009035222

ISBN: 978-81-321-0168-0 (HB)

The SAGE Team: Rekha Natarajan, Sushmita Banerjee and Trinankur Banerjee

This book is dedicated to my six grandchildren, Sonya, Maya, Tabitha, Zinzi, Petra and Milo, who are encouraged to read whatever appeals to their minds and imaginations, think thoughts that uplift and always have a corner in their hearts for their Indian heritage.

Contents

List of Tables and Figures

Tables

Figures

Acknowledgements

Like any book, this could not have been written without the help of many people. However, a special acknowledgement should be made for those who were members of the Arts and Humanities Research Board (AHRB) Research Project on Medieval Kerala Mathematics. In particular, I would like to acknowledge the help of Dr V. Madhukar Mallayya who was the Research Associate and whose expertise and insights were crucial to the successful outcome of the project, especially the comprehensive survey that he carried out for the project entitled 'Trigonometric Sines and Sine Tables in India' (unpublished). This was an important source for Chapter 5 of this book. I would also like to acknowledge the valuable inputs of Mr Dennis Almeida (Member of the research team), Dr M. Vijayalekshmy (Research Assistant), Dr J.M. DeLire (Research Assistant) and Dr C. Goncalves (Research Assistant). Their specific contributions are referred to in various parts of this book.

Note: No mathematical significance should be drawn from the use of lower case and upper case first letters (for example, Sine [sine] or Cosine [cosine]) throughout the book unless specifically pointed out.

1

Introduction

Objectives and Original Features of This Book

The genesis of this book may be traced to an earlier book, *The Crest of the Peacock: Non-European Roots of Mathematics*, first published in 1991. On page 20 of that book I wrote:

> One of the conjectures posed in Chapter 9 (p. 249) is the possibility that mathematics from medieval India, particularly from the southern state of Kerala may have had an impact on European mathematics of the sixteenth and seventeenth centuries. (Joseph 2000)

This single sentence aroused more interest and controversy than other issues raised in the book. At a meeting soon after the book came out, the author was asked whether he was in the business of dethroning Newton and Leibniz! It was also interpreted by some as suggesting that the Indians invented calculus and transmitted the knowledge to Europe. In the course of talks and presentations in conferences presentations in 1990s the author had conjectured that the conduit for such a transmission were the Jesuits whose presence in Kerala from the middle of the sixteenth century was well attested in historical records of the period.

To understand the passions unleashed, we need to stand back and look at the beginnings of modern mathematics. Two powerful tools contributed to its creation in the seventeenth century: the discovery of the general algorithms of calculus and the development and application of infinite series techniques. These two streams of discovery seemed in the early stages to reinforce one another in their simultaneous development by extending the range and application of the other.

The origin of the analysis and derivations of certain infinite series, notably those relating to the arctangent, sine and cosine, was not in Europe, but in an area in South India that now falls within the state of Kerala. From a region covering less than a thousand square kilometres north of Cochin, during the period between the fourteenth and sixteenth centuries there emerged discoveries that anticipate similar works of European mathematicians such as Wallis, James Gregory, Taylor, Newton and Leibniz by about 200 years.[1]

There are several questions relating to the activities of this group (henceforth referred to as the Kerala School[2]), apart from technical ones relating to the mathematical motivation and content of their work. This book proposes to examine both the technical aspects of these mathematical discoveries as well as temporal and socio-economic context of the rise and decline of the Kerala School of Mathematics and Astronomy.[3]

Such a study is timely given the growing interest in recent years in issues such as:

1. the comparative epistemology of Indian Mathematics as distinguished from Western (including Greek) mathematics[4] and
2. the nature of the socio-economic processes that led to the development of scientific knowledge in pre-modern India.

The broad overlapping objectives of the earlier part of this book are:

1. to examine the mathematical genesis of the Kerala School from the earlier history of Indian mathematics in general and the Aryabhatan School in particular;
2. to explore the practices of mathematics/astronomy identifiable as those pursued by the Kerala School;
3. to outline the mathematical content of work done by individual members of the school;
4. to offer possible reasons for the neglect of the work of the Kerala School in standard accounts of the history of modern mathematics;
5. to examine the social origins of the Kerala School through a study of the structure and composition of contemporary society;

6. to explore traditional modes of generating knowledge prevalent in Kerala at the time and in particular how such knowledge was disseminated across space and time and
7. to re-examine the widely-held though mistaken view that Indian mathematics was mainly utilitarian, neglectful of proof and stagnant after the twelfth century in the light of what is now known about Kerala and other medieval Indian mathematics.

The overall objective of the later part of this book is to examine the conjecture of the transmission of Kerala mathematics to Europe, with a view to informing the wider history of mathematics. More specifically, it will examine the following questions:[5]

1. Given the current evidence, what is the extent of the *transmission* of knowledge through Jesuits from Kerala to Europe during the sixteenth and seventeenth centuries? Was Kerala mathematics and astronomy part of this transmission?
2. What was the mode of diffusion of this transmitted knowledge within Europe?

As far as the second part of the book is concerned, there has been little substantive literature on the issue of transmission of mathematics from India to Europe after the twelfth century. Hence, this study will be of deep historical and cultural significance.

Issues Raised and Some Resulting Hypotheses

The mathematicians/astronomers of the Kerala School were predominantly Nambuthiri Brahmins with a few who came from other castes, such as the Variyar and the Pisarati, both traditionally associated with the performance of specific functions in the temple. The personnel of the temple of that time consisted broadly of three groups: scholars (*bhatta*), priests (*santi*) and functionaries (*panimakkal*). The scholars and priests were almost always Brahmins. Of the non-Brahmin functionaries, the Variyars looked after routine tasks of the temple, including keeping the accounts of the temple, and the Pisaratis were a subcaste of priests officiating during the performance of rituals in their own temples. The latter sometimes acted as Sanskrit instructors as well. The two groups seems to have evolved and acquired a high social status over a long period—a period that saw the establishment of the hegemony of Brahmins over the Kerala society through their substantial landholding sanctified by custom (*devasvam*). The Brahmins had the resources and leisure to pursue higher learning, including the study and reinterpretation of certain mathematical and astronomical works of the earlier period from the North. This gives rise to the hypothesis that the genesis of the Kerala School is found in the older traditions prevailing in Kerala and elsewhere in India.

The corporate bodies of Brahmin landholders, organised as temple-centred oligarchies, continued to wield wider control through the creation of new institutional structures such as *yogams* (caste unions) and *sanketams* ('sacred territories' controlled by Brahmins). During the fourteenth century, which marked the start of the Kerala School, there were a number of *yogams* controlled by powerful Brahmin landlords (known as *Nambuthiri-Uralars*). The Nambuthiris were organised along patriarchal lines following a strict primogeniture system of inheritance. In addition to the structures of political importance enhancing their social and economic powers, there was a customary practice called *sambandham*, a form of sexual alliance with the non-Brahmin castes, particularly women from the Nair aristocracy. The eldest son of a Nambuthiri family alone entered into a normal marriage alliance (*veli*) with a Nambuthiri female while his younger brothers, if there were any, could only form a *sambandham* relationship. This arrangement was in a way an interlocking institution of the patriachal Nambuthiri males and matrilineal Nair females. It had the effect of removing any family responsibility from the younger sons among the Nambuthiris while at the same time stabilising the system of matrilineal inheritance among the Nairs.

The system of primogeniture kept the eldest son of the Nambuthiri family busy looking after the property and community affairs while his younger brothers lived unencumbered lives with plenty of leisure time. Lacking in social status and power on a par with the eldest brother, the younger ones needed to attain social respectability through other means. Scholarship, both secular and religious, was one way available to them to make a mark. We therefore formulate the hypothesis that many of the well-known mathematicians/astronomers of the Kerala School may have emerged from among this unencumbered section of the Nambuthiris.[6]

There exist records called *granthavaris* recounting the day-to-day accounts, from the late fifteenth century onwards, of prominent families (*swarupams*), Nambuthiri caste corporations (*yogams* and *sanketams*) and prominent Nair houses (*taravads*). These are in the form of palm leaf manuscripts now being studied by historians for reconstructing the socio-economic history of pre-colonial Kerala. They contain details about economic transactions, social relationships and cultural practices. A noticeable feature is the importance given to the practice of meticulously documenting the events and accounts of the economic transactions of the day.

The system of ownership during the period was not absolute. Even the ruler did not enjoy absolute ownership of land. The ownership of land was a conglomerate of multiple rights and privileges enjoyed hereditarily by different occupational caste groups and organised into a hierarchy. Among these various rights and privileges, the superior kind were vested with the landlords, predominantly the Nambuthiris. The landlords were not free to sell, gift or mortgage their lands to anyone they liked. They could transfer their rights only to equals and the act normally did not affect other rights held by other social groups. Each group could indulge in transactions possible with(in) their entitlements. Commonly, such transactions seem to have been confined to mortgages (*panayam*). The *granthavaris* recorded mortgage of various rights held by the concerned social groups and the deduction of interest out of their entitlements. Since the entitlements varied from a given number/measure of coconuts, areca nuts, paddy and other cereals and pulses, the interest against the cash drawn on mortgage had to be reckoned in terms of inter-commodity exchange ratios. The rights were mortgaged invariably to the landlords and the interest on the cash borrowed on mortgage was deducted out of the entitlement. The landlords thus had to maintain a systematic account of what was given to the subordinates/dependents and the deductions from their entitlements. As evidenced by the *granthavaris*, this accounting was done meticulously. It is obvious that the numerate were skilled in formulaic and tabular devices for deriving multiples and fractions beyond normal comprehension. This should suggest a link between the needs of the numerate with those of the pure mathematicians working at the frontiers of mathematical knowledge and hence a hypothesis about the social origins of higher mathematics in Kerala.

In the history of Kerala's agrarian expansion, the dissemination of calendrical knowledge played a vital role. The knowledge of solar calendar and the skills associated with agrarian management of seasons constituted a crucial source of Brahmin economic domination. Astronomy and mathematics were the two instruments of contemporary seasonal forecasts. The calendar was central to all socio-cultural practices of the period. This again points to the socio-economic relevance of astronomical/higher mathematical learning during the period providing a plausible hypothesis worthy of further examination.

A relevant question in the context of this book that is rarely addressed in studying the historical development of traditional sciences, is how knowledge got generated and transmitted in societies before the introduction of printing and modern schooling system. The traditional societies of Kerala were made up of specialised groups engaged in hereditary trades. It was the responsibility of the elders among each occupation group to train their younger ones in the respective trade. The acquisition of higher knowledge following the teacher–disciple (*guru–sisya*) mode was normally possible only for the elites. For promoting Vedic learning (i.e., learning from the scriptures) there were centres (*salas*), attached to the temples, almost exclusively for the Brahmins. Non-Vedic sciences known as *shastras* were taught by individual scholars at their residence. These sciences, codified in Sanskrit texts, were recorded in palm leaf bundles (*granthas*) that functioned as books. The palm leaf texts were preserved in the houses (*illams*) of scholar teachers who were generally Nambuthiris. Many of the prominent *illams* from the past have yielded Sanskrit manuscripts dealing with epics, grammar, philosophy, astrology and literary compositions. Few texts of higher mathematics and astronomy come down to us. This would lead to the hypothesis that the specialised texts were consulted only by a few. The larger society probably needed only the texts of calendrical calculations and astrological approximations which survived as they were periodically revised and rewritten onto fresh palm leaves by the succeeding scholars.

Approaches to History of Mathematics: Some Reflections

A history of mathematics may be approached in a number of different ways: as a chronological survey; by tracing the development of a particular theme or subject; through exploration of the life and work of individual mathematicians; or by focussing on specific mathematical communities at a particular time and place. The title of this book, *A Passage to Infinity: Medieval Indian Mathematics from Kerala and Its Impact,* could encompass all or some of these different approaches. But the emphasis here is on two main themes: a survey of Kerala mathematics (which is but a continuation of the history of Indian mathematics) and an examination of the conjectured transmission of certain fundamental elements of the calculus from Kerala to Europe through the Jesuits and other conduits.

Now, mathematics is about ideas and their development. It is also about people and societies. Mathematics thus has a history worth knowing and telling. There are also those who feel that the history of mathematics, because it traces the genesis and development of ideas, has a role in education. Both historians and mathematicians have sets of expectations and preconceptions about the history of mathematics that may discourage them from even picking up such a book in the first place. Some may feel that a history of mathematics is little more than a chronology of names of great mathematicians, each associated with one or more theorems or axioms, and perhaps enlivened by an occasional colourful story retold from teacher to pupil down the years with about as much truth content as found in a number of fairy tales. Neither group is knowledgeable about the basic facts and methodology of each other's discipline. The problem then becomes one of persuading these two diverse audiences to pick up a book on the history of mathematics in the first place, and satisfy their expectations enough to keep them reading it. It would be even a greater bonus if one could alter those expectations sufficiently that they find themselves engaged in learning about and enjoying both mathematics and history as they read, without skipping bits concerning 'boring' background or the incomprehensible technical mathematics. For any serious author, there is one dominant standard to satisfy: getting the balance right between the facile and the impenetrable, between the historical context and the mathematical content, without patronising either the readership or compromising one's own intellectual integrity. This is no small task, for if you are reporting on the mathematics of a society from an older era recorded in an obscure language, your chance of success is even more remote. Yet, if you translate the mathematics of that society into the modern symbolic form, you are in the danger of misrepresenting or distorting the original sources and contexts and may well face the wrath of the purist and the pedant. In any case, ancient historians, however engaged they are with the intellectual histories of the cultures they study, have tended to shy away from the more numerate sources of their discipline. Mathematicians, on the other hand, may take one of three or more stances. Some feel that any pre-modern mathematics is trivial and irrelevant to modern mathematics. Indeed, to a number of research mathematicians the history they are concerned with is often three research papers deep. At the other extreme, there are those who already feel ownership of some version of that past through often-repeated legends in the standard textbooks. They may either strongly resent changes to the received wisdom, or simply assume that because it all happened so long ago that there cannot possibly be anything substantively new to tell. The difficult test remains: is it possible to convey the richness, complexity, difficulty and sheer 'otherness' to an audience that one counts not in dozens or even hundreds but perhaps into the thousands and beyond? Whether this book satisfies the test is a matter for the readers and the critics to judge.

Notes

1. Apart from the infinite series for sine, cosine and arctangent there were other basic ideas of calculus discovered by the Kerala School. The most crucial idea often ignored in discussions, was that of asymptotic expansions, perhaps introduced for the first time in mathematics anywhere in the world. Other innovations, expressed in modern terminology, include the solution of transcendental equations by iteration, integration of series term by term, tests of convergence of infinite series and approximation of transcendental numbers by continued fractions. For a useful summary, see Sarma (1972b: 11–28).
2. Madhava of Sangamagramma founded a school that had the following teacher–student lineage:

   Madhava (fl. 1340–1425) ==> *Paramesvara* (fl. 1360–1460) ==> Damodara (fl. 1410–1510) ==> *Nilakantha* (fl. 1443–1560) ==> Citrabhanu (fl. 1475–1550) ====> Narayana (fl. 1500–1575) and *Sankara Variyar* (fl. 1500–1560)

   Also Damodara ==> *Jyesthadeva* (fl. 1500–1610) ==> *Acyuta Pisarati* (fl. 1550–1621)

   The italicized names are generally recognised as the major figures of the Kerala School. Their work will be examined in detail in this book.
3. Discussions of the technical aspects of the 'higher' mathematics are normally found in the appendices and endnotes to relevant chapters of this book.

4. Such a discussion has interesting pedagogical implications as well. The conventional introduction of calculus to students consists of a collection of algebraic techniques that solve essentially geometric problems: calculation of areas and construction of tangents. In the Indian case, ideas of calculus arose essentially as a result of solving algebraic problems involved in evaluating sums and interpolating tables of sines. And this occurred in the absence of the geometrical context found in Europe. The epistemological and methodological differences in the approach to mathematics between the two traditions will be touched on in subsequent chapters of this book.
5. This discussion will be based on the findings of a research project, funded by the Arts and Humanities Research Board, United Kingdom, carried out by the author and Dennis Almeida during the period 2001–2004. [The origins of this collaboration of which the project was the culmination can be traced back to a long discussion between us during July 1997 in South Africa of a possible Jesuit role in the spread of Kerala mathematics to Europe. Previously, the author had raised the issue of the Jesuit conduit at a number of venues, including some in India.]
6. A historical study (Sulloway 1996) in England first published in 1966 and then, interestingly, republished in a new edition in 1996, examined the birth order of leading scientists and philosophers and found that although achievement in conventional terms was generally higher among eldest siblings, truly creative innovations were more often made by those who were younger siblings. For example, Charles Darwin and Benjamin Franklin were the youngest sons of youngest sons for four to five generations. Societal factors would seem in some cases to intensify the effects of birth order. For example, the practice of primogeniture, where the eldest male sibling is the prime inheritee, leads not only to differentiated roles and occupational profiles among siblings but also in the patterns of educational differentiation among siblings. Very little research on the effects of birth order has been done outside western European societies so we do not know how, if at all, cultural and societal institutions influence the psychological factors at work. All we know in relation to Kerala is that a remarkable flowering of mathematical creativity occurred among mainly a group of Nambuthiri Brahmins, the largest local landowning community where an 'extreme' form of primogeniture was practiced. I am grateful to Mary Searle Chatterji for drawing my attention to this point.

2

The Social Origins of the Kerala School*

Introduction

The medieval period of Kerala's history stretches over seven centuries from the age of the Perumals of Mahodayapuram (ninth to twelfth centuries) to the formation and establishment of the *swarupam*[1] organisations (thirteenth to seventeenth centuries). This period was marked by various historical developments, such as the spread of agricultural and village communities, a substantial rise in local as well as overseas trade, and the increasing struggle among the *swarupams* themselves to become the dominant power. It was also during this period that foundations were laid for what came to be known as the Kerala School of mathematics and astronomy.

An agrarian society, found within a monsoon region, needs to know how to predict seasonal and climatic changes. For that an accurate

---

*This chapter is based on two papers entitled 'Indian Mathematical Tradition: The Kerala Dimension' and 'An Intellectual History of Medieval Kerala' which were presented at an International Workshop in Kovalam, Kerala and published in an edited volume of the proceedings as Mallayya and Joseph (2009a) and Vijaylakshmy and Joseph (2009). The author owes a considerable debt to both Drs Mallayya and Vijaylakshmy for their contributions. The reader may find it useful to refer to Table 2.1 and Figure 2.1 in this chapter for both information on the chronology of Indian mathematics and for place names mentioned in this and subsequent chapters.

calendar and astronomical computations of the position and movements of celestial bodies are essential. Earlier work in North India on these topics were studied and translated, and commentaries, written in Sanskrit or Malayalam, elaborated and extended.[2] One such example would be a fourteenth-century text on astronomy by a major figure of the Kerala School, Vatasseri Paramesvara, which became a reference guide for enabling efficient cultivation of both paddy fields and dry land.

Astrology, in Kerala as elsewhere, fulfilled certain needs for finding out auspicious times for specific rituals and observances. Auspicious times needed to be calculated for the naming ceremony, first feeding, initiation, marriage and other *rites de passage.* It was also necessary in the compilation of a horoscope (*jataka*) to foretell the course of life of a person based on the study of the position of certain planets and stars at the time of birth.

The advent of new ideas in medieval Kerala culture came from two sources—from the earlier Indian tradition and from overseas.[3] In the context of this study, the external influence which was mainly confined to trade and settlement of different disasporic groups, would seem to have been relatively unimportant in the intellectual life of the emerging Kerala culture. The Nambuthiris, the Jains and the Buddhists were responsible in different ways and at different times for the introduction from the first source. Sanskrit and other texts containing these traditions originated from different regions of the subcontinent and not all necessarily from the North. During the early part of the medieval period, great importance was given to the study of these texts and attempts made to present the knowledge gathered in regional languages.

The incorporation of new ideas into local cultures affected many disciplines, including literature, religion, art, science and technology and helped the formation of a peculiar regional identity in Kerala. The attempt at borrowing elements of a pan Indian tradition resulted in a surprisingly vast output in scholarly literature. The texts produced were in Sanskrit, in Manipravalam (a mixed language of Sanskrit and Malayalam) and in the emerging new language of Malayalam. Although the elements of this pan Indian tradition had various regional representations, the scientific knowledge emanating from the Sanskritic tradition was confined more or less to the upper strata of society, while its practical aspects (including physical sciences and medicine) became the preserve of the lower strata or of all. Astrology constituted the vocational pursuits of the Ganaka

Table 2.1

Major Personalities and Texts (in Italics) in Indian Mathematics

Baudhayana, Apastamba, Katyayana: 5th–8th centuries BC [authors of *Sulbasutras*]
Bakhshali Manuscript: 2nd/3rd to 7th (?) centuries AD [author(s) unknown]
Aryabhata I: 5th century AD [author of *Aryabhatiya*]
Bhaskara I: 6th century AD [author of *Aryabhatiyabhasya, Laghubhaskariya, Mahabhaskariya*]
Brahmagupta: 6th century AD [author of *Brahmasphutasiddhanta, Khandakadyaka*]
Mahavira: 8th century AD [author of *Ganitasarasamgraha*]
Aryabhata II: 9th century AD [author of *Mahasiddhanta*]
Sridhara: 9th/10th century AD [author of *Trisatika, Patiganita*]
Bhaskara II: 11th century AD [author of *Lilavati, Bijaganita*]
Narayana Pandit: 14th century AD [author of *Ganita Kaumudi*]
Madhava: 14th century AD [author of *Venvaroha, Sphutacandrapti*]
Paramesvara: 14th/15th century AD [author of *Goladipika* and commentaries on *Aryabhatiya* and *Lilavati*]
Nilakantha: 15th/16th century AD [author of *Aryabhatiyabhasya, Tantrasangraha, Golasara*]
Jyesthadeva: 16th century [author of *Yuktibhasa*]
Sankara Variyar: 16th century [author of *Kriyakramakari*]
Putunama Somayaji: 17th century [author of *Karanapaddhati*]
Sankara Varman: 19th century [author of *Sadratnamala*]

community.[4] Medical science was practised among most sections of society, so that for example, there were eminent physicians even among the lower caste Vela or Mannan community.

The Genesis of the Kerala School

The origin of Kerala astronomy is often associated with the legendary fourth-century figure of Vararuci[5] who composed the *Chandravakya* ('moon sentences') to aid astronomical calculations.[6] The *Katapyadi*[7] system of numeration and the *Parahita*[8] system of calculation were also developed around that time as aids to the study of mathematics and astronomy. The *Parahita* system was introduced by Haridatta, in the seventh century as a modification to the Aryabhatan computational system.[9] In his *Grahachakra Nibandhhana* and *Maha-Marga Nibandhhana*, Haridatta used the *Katapyadi* system although it was known by another name until the ninth century when Govindasvamin wrote his commentary (or *bhasya*[10]) on *Mahabhaskariya* of Bhaskara I, a seminal text in the development of Kerala mathematics and astronomy (Unithiri 2003: 47–48). Govindasvamin's text contains some interesting mathematics, including rules for second-order interpolation to estimate intermediate sine values for different intervals, being a special case of the much later general Newton–Gauss interpolation formula.[11]

At the beginning of the ninth century, Kerala underwent certain political changes. The Kulasekharas in the person of Kulasekhara Alvar (AD 800–820) established their power in Kerala. The history of this dynasty has only come to light recently as a result of the study of the inscriptions of that age. It was a dynasty that provided a degree of stabilty for about two centuries and a quarter, starting around AD 900. It was a period of importance in the shaping of Kerala's cultural map, a period of intellectual and creative developments too. The emerging Vedic ritualistic and philosophical scholarship was given a kick-start by Sankara, the famous Advaita philosopher, who was believed to have lived during the early decades of eighth century. Astronomy, astrology, mathematics and health care (*Ayurveda*) received particular attention during the period. A number of palm leaf texts relating to the various forms of knowledge and their manuals of practice vouch for this. Major systems of knowledge in classical India were retold, commented and adapted by the scholars of Kerala. Kautilya's *Arthasastra*, one of the early treatises on state-craft, was recorded in old Malayalam around the twelfth century. *Kutiyattam* and *Kuttu*, the earliest forms of Kerala's dance drama, got standardised and popularised during this period. Sanskrit dramas like *Tapatisamvaranam*, *Subhadradhanajayam* and *Vichchinnabhishekam* were composed and performed during this period. The second in the Kulasekhara dynasty, Sthanu Ravi Varma (844–885), ranks promeniently among the dramatists of the time.

The growing interest in astronomy was given a further boost during the reign of Ravi Varma. He is believed to have established an observatory containing a giant armillary sphere in the capital of his empire, Mahodayapuram (identified as the present-day Kodungallur), under the charge of Sankaranarayana, the court astronomer. A student of Govindasvamin, Sankaranarayana, like his teacher, applied the Aryabhatan planetary model in his calculation, basing it on Bhaskara I's *Laghubhaskariya*. Ravi Varma, also a keen student of astronomy, asked certain insightful questions that were answered by Sankaranarayana in his influential commentary on the *Laghubhaskariya* of Bhaskara I.

The age of the Kulasekharas was notable for progress in the field of education and learning. Schools and colleges attached to temples sprang up in different parts of Kerala. These institutions, known as *sala*s, were maintained with the help of private endowments. Hundreds of youth, including some Buddhists and Jains, were given free board and lodging to study mainly scriptures. In some institutions there were also courses on philosophy (*mimamsa*), grammar (*vyakarana*), law (*dharmasastra*) and astronomy (*jyotisastra*). These *sala*s gradually became an integral part of the temple, as the Buddhist and Jain influences faded, being financed from revenue received by the temple, acting as the landlord, from neighbouring villages and earmarked for this purpose.

Around AD 999, the long period of peace and stability ended with the outbreak of a war between the two major powers of the region. This was accompanied by extensive economic and social changes and the way of life that emerged in Kerala was very different from what it was before. The war called for mobilisation of the resources of the state. The Nambuthiri Brahmins, who had gradually established themselves at the top of the social hierarchy during the earlier centuries, provided valuable financial and other assistance to the rulers. A few of them even gave up their traditional pursuits to take up arms. The growing dominance of the Nambuthiris is shown by two outcomes that would have important implications for the future. First, the power of the monarchy was curtailed, as seen from an inscription found in Kollam (AD 1102) which shows the Nambuthiris in charge of the temple compelling Rama Varma, the last ruler of the Kulasekharas, to donate extensive areas of land to the temple as an act of atonement for some unnamed misdemeanour. Second, it is quite likely that this period marked the beginnings of the iniquitous *Janmi* system of landholding which became the norm over the next 500 years.[12]

Between 1102 and 1498, the period between the end of the end of the Kulasekharas and the coming of the Portuguese, four major states emerged in Kerala, based in Venad, Calicut, Cochin and Kolathunad, with the first two becoming powerful military states. And in the absence of a strong central power, the combined effects of a rigid caste structure, the growing power of the temple and the considerable autonomy enjoyed by local chieftains, led to weak governments in Cochin and Kolathunad. This is particularly relevant to our study since the location of the Kerala

School was mainly in the state of Cochin. To simplify exposition, we will from now on refer to Kerala of this period (i.e., 1102–1498) as 'medieval Kerala'.

For about 300 years after Govindasvamin no major mathematical figure appeared in the Kerala horizon. Indeed, hardly any records exist for sketching developments in mathematics and astronomy from the eleventh to thirteenth centuries. However, it was a period of great activity in North India where mathematician-astronomers, such as Sridhara (c. 900), Aryabhata II (c. 950), Sripati (c. 1040) and the famous Bhaskaracharya (or Bhaskara II, b. 1114), were in the forefront. Some of their work must have slowly percolated into Kerala for they were referred to by those who came later.

The next major figure was Govinda Bhattathiri[13] who lived around 1175. Born in the village of Alathiyur (three kilometres south of Tirur), one of the 32 Brahmin settlements in Kerala established during the seventh and eighth centuries, there is a legend that after performing a *bhajanam* (a special act of worship) to his special deity in the Trichur temple, Govinda was given the gift of foretelling the future. A *swamiyar* (a 'high' priest) who consulted Govinda was told that he had to contend with three additional births, incurred as a result of arousing the anger of Lord Krishna, being in order of their occurrence a rat-snake, a bull and a *tulsi* plant. Possibly as a result, Govinda is said to have left his birthplace for foreign parts (somewhere in present-day Tamil Nadu) to study under a scholar named Kaanchanoor Aazhvar.

Govinda's major work was *Dasadhyaya*, a commentary on the first ten chapters of an astrological text, *Brhajjataka*, by the astronomer, Varahamihira (c. AD 505–587). This is generally recognised as the most important of the 70 known commentaries on this text. A lesser known work of Govinda, *Muhurtaratna*, was referred to by Paramesvara, an important figure in the Kerala School referred to earlier, who described the author as a notable astrologer of his time. Indeed, a famous family of astrologers in Kerala, the Kaniyans of Pazhur, trace their astrological prowess to their supposed descent from Govinda.[14]

Information on the scientific activities during the thirteenth century is hard to come by. We know of a Nambuthiri by the name of Suryadevan who published short commentaries on the works of Aryabhata, Sripati and Varahamihira which would suggest that interest in astronomy continued over the period. However, we had to wait for the next century for the birth of the Kerala School.

The Kerala School: Lives and Works of Its Members

The emergence of the Kerala School of Mathematics and Astronomy may be traced to the early decades of the fourteenth century which saw the appearance of its founder, Sangamagrama Madhava (c.1340–1425). A student–teacher lineage (shown in Figure 2.1) that was to last for more than 200 years serves as a reference point to the discussion of the various members of the Kerala School contained in this section.

Few personal details are known about Madhava. He belonged to a priestly class called the *Emprantiri*, consisting of Brahmins who had recently arrived from coastal Karnataka and become a sub-caste of the Kerala Brahmins. Some information regarding the names of his family house and village are provided in his own composition, *Venvaroha*, in its commentary by Acyuta Pisarati (b. 1550), and in the *Aryabhatiyabhasya* of

Figure 2.1

The Madhava (or Kerala) School of Indian Mathematics

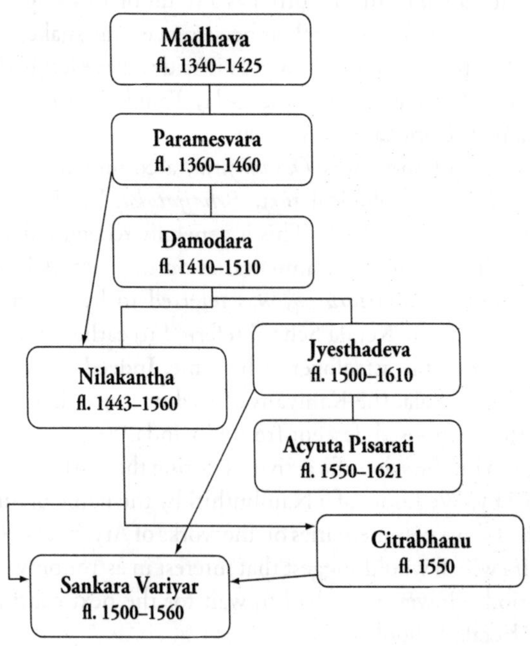

Nilakantha (b.1443). According to this information Madhava belonged to a house called 'Bakuladhistitaviharam' (in Sanskrit) or 'Ilaininnapalli' (in Malayalam) situated in the village of Sangamagrama in central Kerala. This village takes its name from Lord Sangamesvara whose temple is found there. This place is now known by the name Irinjalakuda and situated near present-day Trissur. The 'Ilaininnapalli' of Madhava has been identified as either of two present-day Nambuthiri houses, namely 'Iriarapalli' and 'Iriaravalli', about eight kilometres from Irinjalakudda near Kallettumkara railway station.

Madhava's only surviving works are in astronomy and mainly concerned with refining the *vakya* system of Vararuci, mentioned earlier. In both *Venvaroha* and *Sphutacandrapti*, Madhava carried out a revision of the *Chandra vakyas*, calculating the exact positions of the moon, correct to the second, for every 36 minutes of the day. Before Madhava's work, Varuruci's 'moon-sentences' only gave values correct to the minute. Using the cyclic nature of the lunar *vakyas* where nine anomalistic months equal 248 days, Madhava estimated the lunar longitude at nine equally distant times in one day. He also discussed the computation of the longitudes of the planets and of the ascendant.

Besides these works,[15] it is from a number of stray verses of Madhava quoted by those who came after him,[16] such as Nilakantha, Jyesthadeva, Narayana and Sankara Variyar, that we know of Madhava's original contribution to the development of Kerala mathematics and astronomy. His fame rests on his discovery of the infinite series for circular and trigonometric functions, commonly known as the Gregory series for arctangent, the Leibniz series for π and the Newton power series expansions for sine and cosine correct (in modern terms) to 1/3,600 of a degree. There are also some remarkable approximations, based on the incorporation of 'correction' terms to these slowly converging series that have also been attributed to Madhava. These results will be discussed in detail in subsequent chapters of this book.

Madhava's distinguished pupil was Vatasseri Paramesvara. He was born in Alathiyur in 1360 into a Nambuthiri Brahmin (the Vatasseri) family that specialised in the study of astronomy and astrology. The present-day Alathiyur is situated in the Ponnani taluk of southern Malabar in Palghat district. The location of this village is mentioned in his book, *Goladipika*. Information about his parentage is lacking but from his commentary on Govinda's *Muhurtaratna* and also from his astrological work *Acarasangraha* we can gather information about his grandfather, a well-known astronomer of his time who was also a student of Govinda Bhattathiri who was mentioned earlier. Vatasreni Damodara was Paramesvara's son and an influential teacher who counted among his students Nilakantha and Jyesthadeva, two major figures of the Kerala School.

According to Nilakantha, Paramesvara learnt his mathematics from Madhava. He wrote a number of commentaries, including ones on Aryabhata's *Aryabhatiya*, on Bhaskara I's *Mahabhaskariya* and *Laghubhaskariya*, on *Suryasiddhanta*, on Govindasvamin's commentary on *Mahabhaskariya*, on Manjula's *Laghumanasa* and Bhaskaracharya's *Lilavati*. Paramesvara's role in scrutinising and then disseminating the contents of these major texts of Indian astronomy and mathematics cannot be overestimated. As a result, after Paramesvara, those who followed him had this corpus of knowledge readily available.

Paramesvara's main importance in the development of planetary astronomy in South India is his *Drgganita* system, explained in his text of that name written in 1431. In it he emphasised the importance of checking theories against observations (hence, *drg*: to see or observe; and *ganita*: mathematical theories or calculation). This is an astronomical work in two parts. The first part deals with the derivations of the mean positions and equations of the centre of the planets, the corrections made and the method for calculating the arc from the sine. The second part merely summarises the first part using the *Katapyadi* notation. This was probably the first time in India that a mathematical model of astronomy based on observation was devised. He also wrote *Goladipika* dealing with various aspects of spherical astronomy of which there are two different versions. The two versions together contain a detailed discussion of the great gnomon, shadow and parallex; the construction of an armillary sphere, the apparent and true motions of planets, and methods of measuring the circumference of the earth and other topics.

Between 1393 and 1432, Paramesvara made a series of observations from a nearby seashore of eclipses of the sun and the moon and recorded them. There is a legend that this activity made him an outcaste and so he had to live with low-caste fisherman on the beach to make his observations. This legend is difficult to verify. All we know is that he was a prolific writer and wrote on astrology, including an important commentary, mentioned earlier, on Govinda's *Muhurtaratna*.

The foundation laid by Paramesvara heralded the emergence of the major figure of Nilakantha Somayaji. He was born in 1444 into a Nambuthiri Brahmin family of *Somatris* (i.e., performers of the *soma* sacrifice) in Trikkantiyur in the present-day Ponnani Taluk of the Mallapuram district. His family house has been traced to the present Etamana house occupied by distant relations after the extinction of Nilakantha's family. His younger brother Sankara was also well versed in astronomy. Nilakantha refers to his brother in some of his works. In commenting on Verse 26 in the *Ganitapida* of the *Aryabhatiya*, Nilakantha refers to Sankara's teaching at the house of his patron Netranarayana. Again, at the end of his commentary on the *Goladipika*, Nilakantha states that he is entrusting the commentary to Sankara for its wide dissemination. For his deep knowledge of astronomy, Nilakantha was held in high esteem by his patron and religious head Kausitaki Netranrayana Azhvanceri Tamprakkal. He married Arya and had two sons Rama and Daksinamurti. Both were well versed in *Manusmrti* (Laws of Manu) and other *Dharmasastras* and fluent in three languages—Sanskrit, Malayalam and Tamil. Rama was the author of *Sri Ramayanam* (*Laghuramayanam*).

In *Siddhantadarpana*, Nilakantha gives the names of scholars from whom he had acquired his knowledge. He learnt *Vedanta Shastras* from Ravi, and astronomy and mathematics from Damodara and his *paramguru* (senior teacher), Paramesvara. The notable Malayalam scholar Tuncattu Ezhuttacchan is said to have been taught by Nilakantha. Among his other illustrious students were two members of the Kerala School, Sankara Variyar and Jyesthadeva. Nilakantha wrote a number of influential astronomical works such as the *Golasara*, a short introduction to astronomy, dealing with basic astronomical constants and concepts, the position and movement of planets, the computation of sines, among other topics. His work on the computation of sines is particularly noteworthy and will be discussed in a later chapter. He wrote commentaries and also commentaries on commentaries such as the *Siddhantadarpana* and a detailed commentary on it; an elaborate commentary *Aryabhatiyabhasya* on the *Aryabhatiya* of Aryabhata; the *Chandrachayaganita* and a commentary on it dealing with computations relating to shadows, such as computation of shadow from time (*kramachaya*) and time from shadow (*viparitachaya*). The *Tantrasangraha* is his magnum opus containing several innovative astronomical and mathematical ideas. His *Grahananirnaya* deals with computations relating to solar and lunar eclipses.

Nilakantha's popularity and contacts with contemporary astronomers outside Kerala is evident from his *Sundararajaprasnottara*. This text is an astronomical manual giving detailed answers (*uttaram*) or clarifications to certain astronomical questions or problems (*prasna*) raised and addressed to him by a contemporary Tamil astronomer Sundararaja. Sundararaja's respectful references to Nilakantha as *Sadarsani Paramgata* (one who has learnt the six systems of philosophy) in his *Vakyakarana* is matched by Nilakantha's complimentary reference to Sundararaja's commentary on the *Vakyakarana* in his *Aryabhatiyabhasya*. This dialogue is important because it provides rare evidence of interest in Kerala mathematics and astronomy in other areas of South India.

Nilakantha's fame, however, rests on his *Tantrasangraha*. This work is in eight chapters, containing 432 verses, dealing with various topics connected with astronomical calculations and follows the *Drgganita* system introduced by Paramesvara. It also deals with a variety of subjects including the fixing of the gnomon, calculations of the meridian, of latitude, of the declensions, and so on, and the prediction of eclipses. To illustrate its range and originality, consider briefly two innovations of Nilakantha that have implications for history of astronomy.[17]

In *Tantrasangraha*, Nilakantha carried out a major revision to the Aryabhatan model for the interior planets, Mercury and Venus. In doing so, Nilakantha arrived at a more accurate specification of the equation of the centre for these planets than any other that existed in Islamic or European astronomy before Kepler, who was born about 130 years after Nilakantha. In *Aryabatiyabhasya*, Nilakantha developed a computational scheme for planetary motion which is superior to that of the later one developed by Tycho Brahe in that it correctly takes account of the equation of the centre and latitudinal motion of the interior planets. This computational scheme implies a heliocentric model of planetary motion where the five planets (Mercury, Venus, Mars, Jupiter and Saturn) move in eccentric orbits around the mean Sun which in turn goes round the earth. This model is similar to the one suggested by Brahe when he revised Copernicus's heliocentric model. It is significant that all the astronomers of the Kerala School who followed Nilakantha, including Jyesthadeva, Acyuta Pisarati and Putumana Somayaji, accepted Nilakantha's planetary model. The other works which he wrote in the later part of his life were either commentaries on his earlier works, such as commentaries on *Chandravakyaganita* and *Siddhantadarpana*—the latter is a short work containing 32 verses dealing with important astronomical constants, the theory of epicycles and other matters of topical interest—or works such as *Golasara*, a textbook in three chapters on spherical astronomy mentioned earlier.

It is likely that Nilakantha lived to be over a hundred[18] and during his long life he taught many students, some of them were to become important figures in their own right. His *Tantrasangraha* was a compendium of all the results known up to his time and it generated among those who came after him a number of commentaries, both in Malayalam and Sanskrit, which form an important basis for assessing Kerala mathematics and astronomy.

One of Nilakantha's students was Citrabhanu (fl. 1550). He was a Nambuthiri Brahmin from the Brahmin village of Covvaram (near present-day Trissur). His work, *Karanamrta*, containing four chapters of advanced astronomical calculations within the framework of the *drgganita* system, was composed in 1530. This work also provides the basics for the preparation of the Kerala calendar (*panchagam*). He was also the author of *Ekavimsatiprasnottara* (Twenty-one Questions and Answers), solving each of a set of 21 pairs of simultaneous equations in two unknowns.[19]

A student of Citrabhanu, Narayana (c. 1500–1575), completed one of the major texts of the Kerala School, *Kriyakramakari*, in 1556. A commentary on Bhaskaracharya's *Lilavati*, it was started by Sankara Variyar (AD 1500–1560) a student of both Nilakantha and Citrabhanu. A Variyar[20] by birth, he belonged to the Trkkatiri family based at Trkuttaveli near Ottappalam in South Malabar who were traditionally employed at the local temple. As a disciple of Nilakantha, Citrabhanu and Damodara, Sankara's astronomical lineage could be directly traced to Madhava. A patron of Sankara was Narayana Azhvanceri Tamprakkal, the religious head of Nambuthiris and a great promoter of astronomical and mathematical studies. Sankara wrote a concise commentary called *Laghuvivrti* on the *Tantrasangraha* of Nilakantha at the request of his patron. This work, composed in 1556, is the last work of Sankara. Sankara had earlier written two detailed commentaries entitled *Yuktidipika* and *Kriyakalapa* on the *Tantrasangraha*. However, with hindsight, the most important of his commentaries is the *Kriyakramakari* on the *Lilavati* of Bhaskara II. This work was left unfinished after the verse 199 of the *Lilavati*, but at the insistence of a well-known Kerala literary scholar, Mahisamangalam Sankaran Nambuthiri, his son Mahisamangalam Narayana took up the task of completing the commentary. This commentary is notable in the history of Kerala mathematics and astronomy for its detailed discussion of the works of earlier writers (Govindasvamin, Sridhara, Jayadeva), some of which are not longer extant, and for providing a rationale and proof of a number of earlier results. Two other works attributed to Sankara are the *Karanasara* and an elaborate commentary on it in Malayalam, namely the *Karanasarakriyakrama*. The *Karanasara* is an astronomical text in four chapters composed in 1550.

Another student of Nilakantha and Damodara and a contemporary of Sankara was a Nambuthiri called Jyesthadeva (fl. 1500–1610), the author of a critically important text of the Kerala School called *Yuktibhasa*. He belonged to the Paraottu family from the Alathiyur village, the birthplace of some of the members of the Kerala School mentioned earlier. The Paraottu house still exists in Alathiyur.

The *Yuktibhasa* is in two parts. The first part is in Malayalam and it deals with rationale of several mathematical results of great significance. It also contains several results of Madhava pertaining to infinite series for π and sines, computations of sine values, series approximations along with their detailed rationale.[21] The second part of the *Yuktibhasa* is an astronomical treatise giving the rationale of various astronomical results. C.M. Whish, the first Westerner to take note of the work of the Kerala School, ascribes to Jyesthadeva the authorship of another astronomical treatise *Drkkarana* in Malayalam composed in 1608. This text is no longer extant.

There are at least three versions of the *Yuktibhasa* of which the Malayalam version became an important source of dissemination throughout Kerala.[22] Based on Nilakantha's *Tantrasangraha*, it is unique in that it gives detailed rationale, proofs or derivations of many theorems and formulae in use among the astronomers/mathematicians of that time. This text will be referred to a number of times in the subsequent chapters.

Jyesthadeva's disciple Acyuta Pisarati refers to his teacher in reverential terms in his work *Uparagakriyakarma* and indicates that Jyesthadeva was very old at the time of composition of the work in 1592. A Malayalam commentary confirms that Acyuta's work was based on his interpretations of the instructions that he had received from Jyesthadeva.

Acyuta Pisarati came from a community who were not Brahmins but performed traditional functions as cleaners and suppliers of flowers and plants for the temple. They were also employed by some Nambuthiri families to give advanced instructions on the calculation of the astrological calendar (*panchagam*) and time reckoning.[23] He lived during 1550–1621 and also came from the epicentre of Kerala mathematics—Trkkantiyur near Tirur in the Ponnani Taluk of south Malabar.

Acyuta was a considerable scholar who made a mark not only in astronomy but in literature and medicine. He attracted the attention of Raja Ravi Varma of Venad and earned the highest praise for his astronomical skills such that a contemporary, Vasudeva, described him as greater than even Lord Siva! His major contribution is found in his work, *Sputanarnaya* (composed before 1593), where he introduces for the first time in Indian astronomy, a correction now known as the 'Reduction to the ecliptic', around the same time as Tycho Brahe did in Western astronomy.[24] A rationale for *Sputanarnaya* was provided in the *Rasigolasphutiti.*

In other works,[25] Acyuta showed his considerable versatility, by not only composing standard treatises in astronomy, but also astrological works such as the *Jatakabharanapaddhati* and the *Horasaroccaya* based on Sripati's *Jatakapaddhati.* He also wrote a Malayalam commentary on Madhava's *Venvaroha.* His students include scholars like Melpattur Narayana Bhatta (a poet and grammarian) and Trpanikkara Poduval (an exponent of *Jyotisa*).

In Charles Whish's 1832 paper, which drew the attention of the world for the first time to the existence of Kerala mathematics and astronomy, appears the passage: 'The author of the *Karanapaddhati* whose grandson is now alive in his 70th year was Putumana Somayajin, a Nambutiri Brahmana of Trisivapur (Trichur) in Malabar.' A major work in the dissemination of Kerala mathematics and astronomy not only in Kerala but also in the neighbouring areas of present-day Tamil Nadu and Andhra Pradesh, *Karanapaddhati* was recorded in 1732, about 200 years after Jyesthadeva's *Yuktibhasa.* The author belonged to the Putumana family of Nambuthiris and came from the village of Covvaram near Trichur where a house by the name Putumana of traditional astronomers still exists.

Karanapaddhati is a comprehensive treatise covering Kerala mathematics and astronomy. It has one unusual feature. It follows generally the *Parahita* system and only advocates the *Drgganita* system in the calculation of eclipses. In ten chapters it discusses problems that appear in earlier texts, like the *kuttakara* approach to solving indeterminate equations, or the derivation of implicit values for π and for sines and cosines of angles. While it covers more or less the same ground as the *Yuktibhasa*, its non-technical clarity in explaining from the first principles methods of deriving various formulae and construction of tables of astronomical constants meant that it became an important source for commentaries, with two in Malayalam, two in Tamil and one in Sanskrit having been discovered so far. Putumana Somayaji also wrote an elementary manual, *Nyayaratna*, for explaining 'astronomical rationale to the dull-witted' and practical texts such as *Venvarohastaka* for determining the positions of the moon at regular intervals.

After Acyuta, little in the way of original work was done, although the tradition of providing corrections and contributing to the preparation of the astronomical ephemeris for the daily needs of the faithful observers of *muhurtha* and practitioners of *jataka* continued for a long time. The compilation of the *panchagam* (calendar) was periodically subjected to *sphuta* (or refinement). About 100 years after *Karanapaddhati* appeared the last of the known texts of the Kerala School. The author of this book, Sankar Varman of Katattanad in north Kerala belonged to the royal family of that area and was a contemporary of Charles Whish. His book, *Sadratnamala*, written in 1823, contains many of the results of the Kerala School, given without the rationale or derivations found in the earlier texts. Whish met him and described him as 'a very intelligent man and acute mathematician'. He died six years after Whish's article on Kerala mathematics and astronomy appeared in 1832.

This is only a short account of a vast tradition and as such only a few landmarks on the highway have been touched. Explorative studies have been carried out only on a small percentage of the mass of manuscripts that have come down to us from the past. An enormous amount of primary material lies unexplored in various repositories. In a monograph entitled *Science Texts in Sanskrit in the Manuscripts Repositories of Kerala and Tamilnadu*, K.V. Sarma (2002) identified as many as 3,473 science texts in Sanskrit and 12,244 science manuscripts from more than 400 repositories in Kerala and Tamil Nadu. Many of these have not been catalogued nor subjected to critical scrutiny. This remains a task for the future.

Nambuthiri Brahmins and Medieval Kerala

The origin of the Nambuthiri Brahmins in Kerala is buried in myths and legends. There is a popular folklore that the legendary founder of Kerala, Lord Parasurama, finding nowhere to go after being banished, obtained the permission of Varuna, the God of the Sea, to reclaim from the Indian Ocean all land within the reach of a throw of his axe. With a mighty heave, Parasurama threw his axe from Kanya Kumari, the tip of the Indian subcontinent, to Gokarnam, which was then buried under the sea. The sea receded and Kerala came into existence.[26] To populate this land, the legend continues, Parasurama invited a group of Brahmins from the north and gave them the land. He also brought a number of dependents to serve the Brahmins. For them, the Nairs, he instituted the *marumakkattayam* system (a system of inheritance and descent through the female line) without the formal institution of marriage. *Sambandham* (or concubinage) between Nair women and Nambuthiri men was permitted, and indeed in some cases even encouraged, with children of such unions becoming the sole responsibility of their mother's family. At the same time, the Nambuthiris followed a system of patrilineal descent (*makkatayam*), with an unusual form of primogeniture that allowed only the eldest son to inherit land and property and marry Nambuthiri women up to three.[27] Only in the case of a lack of a male heir on the part of the eldest son could a younger brother take a wife from the Nambuthiri women. The eldest son was also required to provide for the material needs of his siblings.[28]

This legend may be little more than an attempt at *a posteriori* justification of some of the singular features of the traditional society in Kerala: the economic and social dominance of the Nambuthiris, the matrilineal system of Nairs and their close ties with the Nambuthiris, including their duty to serve and protect them.[29] By implication, this legend reinforced the convention that the 'sacred' duty of the Nairs was to protect the Nambuthiris from lower castes who were the rest of the population, with the exception of Syrian Christians and Muslims who enjoyed special privileges as outsiders. And this protection was achieved through an elaborate system of caste hierarchy and ritual pollution. Indeed, apart from the Muslims and Christian populations who were 'pollution neutral', the rest of the population (the *avarnas*) were put on a hierarchy of pollution proneness depending on their permitted distance of approach to the high castes, consisting mainly of Nambuthiris and Nairs (the *savarnas*).[30]

Perhaps, what the legend confirms is that at a certain time in the distant past a group of Nambuthiri migrants moved south and proceeded to settle in 64 'villages' (or *gramams*) in a region which now lies between south Karnataka, where *Tulu* Brahmin settlements were already in existence, to central Travancore where settlements began appearing before the end of the eighth century.[31]

Irrespective of when they first entered Kerala, the Nambuthiris soon made their presence felt. Their religious authority and reputed scholarship brought them to the attention of the rulers and soon established an influential position in the courts. It was perhaps the Nambuthiris who introduced the mainly two-tier caste system in Kerala with the Nairs being given the status of *Sudra*, the fourth in the conventional caste hierarchy after *Brahmin*, *Kshatriya* and *Vaisya*. The legend that all lands rightfully belonged to Nambuthiris having been given to them by Parasurama gained wide currency and this helped them to acquire vast tracts of land and establish the *Janmi* system of land-holding. However, they were shrewd enough to recognise that an authority based on religion and legend could prove to be temporary. Hence, there emerged the institution of the temple as a surrogate landlord owning large tracts of land with the control and management of the revenues received being in the hands of the Nambuthiris.

The Temple Culture of Kerala

The Nambuthiri Brahmin settlements contained a peculiar temple-centred organisation, consisting of a central temple (*gramaksetra*) surrounded by a periphery of subsidiary temples. The administration of the temple and its property was controlled by a synod of wardens (*urnma*) or council, consisting of hereditary representative of founding families.[32] From the records of land grants during the eleventh and twelfth centuries, these founding families were those Brahmins who had been invited by a ruler when the temple was first built to settle around it, who then proceeded to manage it and its properties.

Under the control of Nambuthiri Brahmins the temple became a device through which they exercised both spiritual and temporal power. The council or its constituent bodies met regularly in the precincts of the temple. Important decisions (usually unanimous for any dissident would be excluded and ostracised) taken by the council were often written on granite walls or slabs or more unusually copper plates. In many cases, the local ruler was invited to preside over the council and was entrusted with the implementation of council decisions. Generous endowments by various sections of the population made these temples immensely rich.[33]

The concentration of agrarian activities under the supervision of the temple resulted in the establishment of an elaborate social order and an unprecedented expansion of agriculture. The resources of the temple were so considerable that it took the lead in carrying out programmes for raising agricultural productivity such as initiating large irrigation projects and reclaiming water-logged land for cultivation.

It would appear that by about the eleventh century most of the forested valleys of the eastern regions of Kerala and the fertile highlands in the interior were brought under cultivation by temple corporations. The temple corporations had to give certain dues to the royal authorities who protected the temple and its land. An interesting feature was that a major share of the revenue received by the royal authority was returned as donations to the temple for some ritual or ceremonial purposes.

The temple employed a number of persons in various capacities. As the largest landowner, it offered employment opportunities for hundreds of persons, some of whom received in return allotments of land as hereditary right (*vrutti*) and others right to land for life (*jivilam*), with the former coming from groups such as priests, accountants, dancers, musicians and teachers and the latter consisting of manual workers such as cleaners, sweepers and watchmen. Several influential functionaries such as the *variyan* (accountant) or *potuval* (secretary) concerned with the management of temple property and services and remunerated through land allocation gave birth to small sub-castes known as *ambalavasi* (temple servant) or *antaralajati* (intermediate caste). Probably as a result of their close association with the Brahmins in the temple, there emerged among them scholars of Sanskrit, physicians, astronomers and poets who were to rival the Nambuthiris themselves.

The development of various land tenure systems through the redistribution of resources of the temple went hand-in-hand with the formation of power groups in the temple. First, there was the *subhayar* (or members of a *sabha* or an assembly) who were both custodians of the temple and owners of large landholdings. They enjoyed a high socio-ritual status coming, as they did, from the caste of scholars and priests. An executive subcommittee formed from the *sabha*, known as the *paratai*, ran the temple affairs for a fixed period of office. The *sabha* also organised themselves into *ganams* (or trusts) to look after endowments involving landed property.[34]

The personnel of the temple were divided, as mentioned earlier, into three groups: scholars and priests and functionaries. The scholars and priests were always Brahmins. Among them were *agamic* instructors, *somatris*, teachers (*bhattas*) and students (*cattinars*). The students were Brahmin youth in *salas* who were expected to protect the temples at times of strife.

Of the non-Brahmin functionaries (or *ambalavasis*) was the *potuval* (literal meaning 'public servant') who received all endowments and donations on behalf of the temple corporation. Then there were the Variyars and the Pisaratis whose functions were described earlier. Drummers, dancers, musicians, these were a substantial group of temple functionaries who were not Brahmins. There were often one or more temple dancing maids or courtesans attached to a temple. There were also the manual workers such as sweepers, suppliers of firewood, suppliers of banana leaves to eat on and gate keepers.

The leaseholders for the land owned by the temple were mainly non-Brahmins and included Christians and Muslims. An important group of them was a sub-caste of Nairs, the progeny of Nair women and Nambuthiri males. They became the intermediaries in transactions between the temple and its tenants to emerge as a powerful group. From the rents extracted from the tenants, they paid their dues to the temple and retained the surplus for their use. As a result, they became highly prosperous and from their ranks came both local chieftains (*samantas*) and military leaders.

The role of the temple as an institution creating and sustaining intellectual activities needs further investigation. As a meeting place of those involved in the study of mathematics and astronomy, as a vehicle for receiving and disseminating scientific knowledge, as an agency for recruiting able students and practitioners outside the confines of narrow caste and regional lines, in all these cases the temple may well have played an important part. The specific questions that deserve to be addressed in the context of this study include the following:

1. How did non-Brahmins such as Sankara Variyar and Acyuta Pisarati become important members of the Kerala School?
2. What was the nature of the relationship between the Tamil astronomers, such as Sundararaja, and members of the Kerala School, such as Nilakantha, as exemplified by the book *Sundararaja-prasnottara* (which, as mentioned earlier, was in the form of questions posed by the former and answered by the latter)?

The temple was the principal but not the sole institution through which the Brahmins exerted their influence in medieval Kerala. In an age when religion was inextricably mixed with politics the suggestion and advice offered by religious leaders were rarely ignored by the rulers. The Brahmins were employed in various capacities: ministers of the king, officers of the court, military leaders in the field, but above all as spiritual preceptors to turn to for advice on personal or state matters. The ubiquitous role of the Brahmin is well illustrated by the fact it was one of them from the court of the Raja of Cochin who led Vasco da Gama to the Zamorin of Calicut in 1498. It was also a Brahmin who was employed by the Zamorin to report on the activities of Vasco da Gama at the court in Cochin. No territorial boundaries seem to constrain the Nambuthiri Brahmin. A Brahmin employed by the Cochin Raja had as much liberty to visit the Zamorin's palace and enjoy his hospitality even during the height of hostility between the two rulers and vice versa. Sharply divided on all questions, the Zamorin and the Raja of Cochin were one in their great respect for the Brahmins. This was of course not peculiar to Kerala, but in no other part of India did the Brahmins possess so much secular and political authority. Like the Catholic Church in medieval Europe, the Nambuthiris became a supranational body owing only nominal allegiance to the secular power.

Another factor that contributed to the enormous influence of Brahmins on political life was their alliance with the monarchy and the leading Nair families. As mentioned earlier, according to the custom among the Nambuthiris, only the eldest son was allowed to marry from his own caste. Therefore, the younger members sought alliances with royal and high caste Nair families. Many of the wealthy Nambuthiris were related to a number of such families and this enabled them to wield considerable influence on the political and religious affairs of the day.

There were also instances of long lived dynasties of Nambuthiri rulers of small kingdoms such as Edapalli and Ambalapula who had an influence disproportionate to their size. A good illustration is the tiny kingdom of Edapalli whose Raja was always a Nambuthiri Brahmin for many centuries. The sacerdotal character of the ruler of this kingdom was probably responsible for its unbroken independence during the period of study. It was the place which usually gave asylum to the Cochin Raja whenever he was defeated by the Zamorin. Even the ruler of Travancore, who had annexed territories that belonged to the Cochin Raja, spared Edapalli because of its Brahmin ruler. It was a kind of asylum, like a free town or a cathedral in medieval Europe, to which people who were afraid of persecution or punishment could retreat to for safety.

The Nambuthiris laid down rigid custom-prescribed rules of life regulating even their and others most trivial actions. The Nairs were prohibited from learning Sanskrit in their *kalaris* so that all the learned professions came to be monopolised by the Brahmins.[35] The rigid caste observances they followed and the position they occupied also helped them to maintain their dominance for many centuries.

The appearance of Vasco da Gama in 1498 was an event that was to alter the whole course of Indian history. The conflict between the Zamorin of Calicut and the Raja of Cochin was accentuated after the former refused to give the Portuguese a monopoly of pepper trade and the latter becoming a Portuguese protectorate. The Portuguese dominance was of short duration and was replaced by the Dutch and by the end of the eighteenth century by the English.

However, there were two domestic events that had extensive effects on the Kerala society. In 1729, Marthanda Varma of Travancore ascended the throne and went on to establish his authority by curbing the powers of the Nambuthiris and the local chieftans. About 60 years later, Saktan Thamburan did the same in Cochin. The invasion of the Mysore sultans in the latter half of the eighteenth century hastened the decline of the Nambuthiri dominance since this privileged group bore the brunt of the anger of the sultans. The old feudal and caste-ridden society of Kerala was undermined never to recover its former position. At the turn of the nineteenth century, the establishment of the British suzerainty over what came to be known as the princely states of Cochin and Travancore and the annexation of Malabar to become a part of British India hastened the process of disintegration of the traditional structure and dealt a fatal blow to the power of the Nambuthiris.

Knowledge Acquisition in Medieval Kerala

Introduction

An important question rarely addressed in history of the sciences is how people acquired knowledge before printing and establishment of modern schooling. It is difficult for us with the easy access to books, the internet and other sources of information to imagine the situation in societies lacking these essential prerequisites. But this is an important question for which an answer is required to obtain a fuller understanding of the mechanics of acquiring and restricting the flow of knowledge.

There is a short answer to the question. Learning was mainly through personal contact and traditional Kerala culture was organised in such a manner to facilitate such contact. It is important in talking about traditional culture that one distinguishes between its two facets: 'popular' and 'high' cultures. In the former, knowledge is passed on not by self-conscious instruction arranged at an institutional set-up for that purpose, but through continuous and customary interpersonal contacts within the family and immediate social group. In the latter, working within the limits of what is possible, a few chosen individuals (usually males from the higher castes) are instructed on the arts, the crafts and the sciences of that culture. Knowledge passes directly from teacher to student and from generation to generation, the emphasis being less on comprehension but the ability to render faithfully the knowledge acquired. In such a system the dissemination of knowledge was *formal* and *intensive*.[36]

Imparting traditional knowledge was a time-consuming process. Early in life, a child was trained to accept without questioning prescribed forms of knowledge, dictated by custom and convention. Accurate repetition of memorised passages and strictly prescribed performance of tasks went hand in hand with respect for authority.

Thus, there existed a system of knowledge acquisition and transmission which was both authoritarian and hierarchical, cultivating an attitude of respect to cultural norms and socially elevated persons as embodiments of status and value. The cultivation of respect was achieved at the social level by the practice of disciplined obedience, symbolically exhibited by ceremonial and ritual deference, and at a personal level by subscribing to ideals of devotion towards elders.

The Mechanics of Imparting Traditional Knowledge in Kerala

In both the practical and intellectual disciplines within traditional Kerala culture, rituals played a central role. In agriculture, trade, statecraft, martial and performing arts, fashion and design, ornamental arts of carving and pottery, manufacturing arts of metal work and textiles, and construction art of architecture there was a ritual dimension to all these activities. It was rituals that symbolically authenticated the traditional forms of these arts by invoking natural and supernatural powers.

The traditional sciences (known as the *shastras* after the classical texts in which they were codified) were taught through recitation of verses or sometimes prose aphorisms, both of which were seen as suitable medium for easy memorisation and contemplation. This was supplemented in more advanced study by reading commentaries of these texts, together with oral discourse with the teacher. Thus, like a number of other aspects of traditional culture, science was presented through forms that were highly intensive in character. There was only a limited amount that could be memorised and hence the highly condensed form of the verses and aphorisms, often cryptic to the point of incomprehension. They required considerable interpretation and explanation but were also intended to lead the mind to deeper understanding through repetition and thought.

Like the scientific disciplines today, traditional disciplines could be either about techniques and their application or theories and their understanding. In either case, in the traditional cultures, they were mainly empirical in the sense that they originated or were based on practical experience. But certain circumstances required these discipline to be presented in a certain way: in a concise form consisting of authoritative assertions and appeal to established principles. It was often left to the student with the help of the teacher to relate these instructions to practical experience. It was therefore unprofitable to search in a traditional text, which followed a pithy presentation, for derivation, explanation and justification.

Thus, the teacher–student relationship was crucial in the dissemination process and the personal bond between teacher and student was valued above all else. During the period under study there prevailed the *Gurukula* system of education.[37] Under this system, a student after completing his early education at home was accepted into a *Gurukala* after evaluation by the Guru. Once accepted, the student became a member of Guru's family. The student would then follow the instructions of the Guru, both serving and obeying him implicitly. This learning process involved three stages: learning from the teacher through oral instruction leading to a Guru–student bond (*gurusishya parambara*); learning through participation in discussions (*parishads*) where the great masters presented and discussed their findings; and through a type of meeting (*vidvat sadas*) where scholars presented effective arguments and counter arguments on various issues.

The celebrated *Gurukulas* of Kerala were the Trichur Brahmaswam Matham, Thirunavaya Samuha Matham, Kudallur Mana Gurukulam, Thiruvalla sala, Moozhikulam sala and later Kudungallur and Punnasseri Gurukulam. Thiruvalla was noted for instructions in arms to the Nambuthiri students. The Kudallur Mana Gurukulam was noted for learning in grammar. There is a shrine of the great Indian grammarian, Patanjali, at Kudallur Mana, perhaps the only shrine of Patanjali that survives in India today. It should be noted that in the *Gurukula* boarding and lodging and tuition were free. Simple living and high thinking was greatly encouraged for it was believed that greed and pursuit of wealth would lead to self-destruction.

Contemporary literary competitions like *Revathipattathanam* at Kozhikode and *Kadavallur Anyonyam* at Kadavallur (near Kunnamkulam) were evidence of the literary and intellectual liveliness of the period. Students from the *Gurukulas* mentioned earlier were expected to participate in these competitions. Uddanda Satrikal, the famous Sanskrit poet, once visited Kudallur Gurukulam and was highly impressed by the standard of the curriculum and the tradition of deep scholarship found there.

THE *ILLAM* AS AN EDUCATIONAL INSTITUTION

In medieval Kerala, people lived in scattered settlements. Their contacts with outside world was through travel and hence learning was acquired in three main ways. First, scholars were invited to certain cultural centres. These centres were established or patronised by notables, often of royal birth, some of whom were considerable scholars or artists in their own right. Second, in certain temples, rich people donated part of their wealth to art and learning as an expression of their piety. Part of this donation was used to support scholarly activities. The temple was also a place of pilgrimage where people travelled long distances to see the divine in different forms: in sculpture, architecture and ceremonial drama or in priests and scholars given to a life of devotion or cultivation of intellectual pursuits.

Third, and more commonly in the case of the Nambuthiris, there were homes (*illams*) of individual teachers whose fame attracted students from far and near to stay for extended periods of time.[38] This type of transmission was important in a society where knowledge was passed on from person to person without recourse to books or similar aids.

It was the practice among the students and teachers of the Kerala School to live and work in these *illams*. The typical *illam* had one or two internal courtyards without a roof and was built either as a square or double square around them. The roof was tiled, or more commonly in the olden days thatched with coconut leaves. The thatched roof had to be replaced annually. The floor was bare, just constructed from mud: no form of cement or marble was used. There would be a shed some distance away for the cows and a sizeable stack of hay to feed them.

Membership of a number of these *illams* were often limited, as we saw earlier, to a leisured class of younger sons of land-owning families whose economic needs were provided for by their families and whom custom dictated could not marry. A few of them chose to take up the study of astronomy and mathematics as an intellectual activity and not for earning their livelihood by constructing calendars or casting horoscopes. The latter was generally undertaken by another group consisting of professional astrologers (*kaniyans*) whose technical knowledge of these areas rarely went beyond a minimum basic competence passed on through families and through referring occasionally to certain elementary manuals. However, a symbiotic relationship apparently developed between some members of the Kerala School and certain *kaniyans*, transcending caste barriers, providing a stronger 'application' orientation to the work of the former group and improving the mathematical and astronomical skills of the latter. It is interesting how, even in a highly rigid caste system, contacts were maintained between two groups who were socially far apart, raising the wider question of the extent to which scientific ideas percolated into the wider society.

In a number of *illams* which specialised in mathematical astronomy, a small, self-selected and highly motivated group of students were set to memorise verses summarising results from the past. It is from the works of these scholars that we can piece together the Kerala episode of Indian mathematics. And for the most part, the writings of these scholars rarely circulated much beyond their locality. Contrast this with the wider campus within which the proliferation of mathematical discoveries occurred in Europe during the seventeenth and eighteenth centuries and perhaps we have some clue as to the different course of mathematical development in the two cultures.

Conclusion

Between fourteenth and seventeenth centuries, when Kerala mathematics and astronomy was in the forefront of all such work in India, a group of Nambuthiri Brahmins, many of them younger sons whose material requirements were provided for by their families and whose family responsibilities were minimal, had at their disposal considerable leisure time to pursue various activities. Some studied religious texts, others wrote erotic poetry, and a few engaged with a single-minded devotion their interests in scientific pursuits. It was in such a social setting that Kerala mathematics and astronomy was created and sustained for over 500 years.

Notes

1. *Swarupams* were ruling families which came to control the Nadu divisions. They were large extended families whose political authority was organised on *kuru* (i.e., order of seniority). For further details, see Raghava Varier (2002).

2. About 400 palm leaf manuscripts on astronomy and 350 on astrology have been discovered in Kerala. Many remain hidden from public view due to the conservative attitude of the custodians of these manuscripts. In the hot and wet climate of Kerala a number have perished or are in such a poor state that it is impossible to decipher them. Certain areas of Kerala, especially in Malappuram and Trichur districts, were the centres of mathematical and astronomical activities during the medieval period. These areas have not been sufficiently trawled for new sources of manuscripts. (See the various 'censuses' of K.V. Sarma, listed in the Bibliography, starting with his *History of Kerala School of Hindu Astronomy* [Sarma 1972].)
3. The influence of alien cultures coming in the wake of overseas trade led to the formation of trade and cultural diasporas along Kerala coast. The settlement was a window to foreign culture where foreign culture was manifested through food, dress, trade mechanism, and so on. For details of trade diasporas, see Curtin (1984: 10–11).
4. Some of the *ganakas* (or traditional astrologers who came from a low caste group) were highly respected during this period in providing auspicious times for holding court festivals, undertaking state activities and even declaring wars. For example, one of the most popular texts was *Rana Deepika* by Kumaraganaka (14th–15th century) who enjoyed the patronage of Deva Sarma a prince of Ambalappuzha Royal family. For further details, see Ulloor Paramesvara Iyyer (1953: Vol II, pp. 112–13).
5. A number of Vararucis make fleeting appearances in different centuries.The author of one of the *Sulbasutras*, Katyayana, was also known as Vararuci in Kerala which would place him around 500–800 BC! An important personage in the court of Vikramaditya was called Vararuci which would mean that he lived about AD 200. One of the major ancient astrology texts called *Vararuci Keralam* was work of yet another Vararuci. A popular legend has it that this Vararuci married a Parayan (low caste) woman and had 12 children by her, each of the children being brought up by the family of a different caste. The eldest child is supposed to have lived between AD 343 and AD 378. This would be the most likely Vararuci referred to in contemporary literature.
6. Through a series of 'nonsense' mnemonic words or phrases, the positions of the moon at regular intervals each day was traced to help worshippers carry out their daily observances and rituals.
7. The *Katapyadi* was a refinement of an earlier system of numerical notation, the Aryabhatan alphabet-numeral system to be discussed in the next chapter. In *Katapyadi*, the Sanskrit letters *k* to *j* indicate 1 to 9, and so do *t* to *dh*; *p* to *m* stand for 1 to 5 and *y* to *h* for 1 to 8. A vowel not preceded by a consonant stands for zero. In case of conjunct consonants, only the last consonant has a numerical value. The number–words are read from right to left so that the letter denoting the units is given first and so on. If such a system is applied to English, the letter *b, c, d, f, g, h, j, k, l, m* would represent the numbers 0 to 9. So would *n, p, q, r, s, t, v, w, x, y*. Let the remaining letter *z* represent 0. The vowels *a; e, i, o, u* serve the function of helping to form meaningful words. Thus the old name for the South Indian city Chennai, 'Madras', would be represented by 9234 and the word 'love' by 86 read in the usual manner from left to right. A system devised to facilitate memorisation, since for any particular

number, different memorable words could be made up with different chronograms. For further details on the origin of the *Katapyadi* system of numerical notation, see Madhavan (1991).

8. *Parahita* literally means 'desired by others' or 'suited to others'. The system is a description of certain corrections to the Aryabhatan planetary system introduced by Haridatta. The *Parahita* system became the cornerstone in the propagation and practice of astronomy in Kerala.
9. Haridatta also devised the system of constructing graded tables of the sines of arcs of anomaly (*manda-jya*) and of conjugation (*sighra-jya*) at intervals of $3^\circ 45'$ to facilitate the computation of the true positions of the planets. It is important to note in this context that Indian astronomers, especially after Aryabhata, were driven by the need to develop the 'best' algorithms, for they noted that, over a period of time, discrepancies between computation and observation tended to increase. Explicit statements were therefore made about the need for astronomers to sit together and decide how the algorithms were to be modified or revised to bring computation back into agreement with observations. According to a popular account, an assembly of astronomers was called at Tirunavaya on the banks of river Bharatapuzha in northern Kerala on the occasion of the 12-yearly 28 days festival called *Mamannka mahotsavam*, to discuss the shortcomings of the prevailing astronomical methods of Aryabhata and his school and to reform the system by simplifying the Aryabhatan methods. For further details, see Ulloor Paramesvara Aiyar (1953: Vol 1, p. 165).
10. A *bhasya* is more than a commentary, for apart from an exhaustive survey of relevant literature and a study of the text, it often served as a vehicle to initiate original investigation on the same and related topics.
11. Govindasvamin's procedure is an extension of Brahmagupta's interpolation procedure discussed in his book *Khandakadyaka* (AD 665). For further details, see Gupta (1969: 86–98).
12. During the period of the war, the Nambuthiri Brahmins became trustees of substantial properties owned by temples. The temple wealth was further augmented by transfers of land and other endowments belonging to individuals who did so as an insurance against devastation or to obtain exemption from taxation. It was under such circumstances that the *Janmi* system originated in Kerala. The *Janmi*, who was in almost all cases a Nambuthiri landlord, exercised not only absolute proprietorship of land and the resulting powers to evict tenants from his land at will but even the power of life and death over them. A system of agrarian serfdom came into existence where the sale of any land meant that its tenants and workers followed as chattels.
13. The Bhatattatires, a sub-group of Nambuthiri Brahmins, became the custodians of Vedic knowledge. They performed all the duties associated with the sacred fires (*agnihotris*), except the performance of sacrifices. After an intensive study of the Vedas, they were required to learn logic (*tarka*), religious philosophy (*vedanta*), grammar (*vyakarana*) and rituals (*mimasa*) and then teach these disciplines at the *salas* attached to the temples.
14. This connection is supposed to date back to a night which he spent with a woman who belonged to the Kaniyan family at Pazhur in the Vaikom Taluk. The progeny of their union is believed to be the ancestor of this famous family of astrologers.

15. There have been other minor works, available only as manuscripts, namely *Madhyamanayanaprakara*, *Mahajyanayanaprakara*, *Lagnaprakarana*, *Aganitagrahacara* and *Aganitapanchanga*.
16. The attribution to Madhava usually took the form of either a statement such as '... *ata eva Madhavopyaha...*' ('... hence Madhava said...') preceding the quotation of a result or a statement such as '... *Iti jyacapayah karya grahanam Madhavotitam*' ('... computation of the arc from the sine and cosine is given by Madhava'). There is every likelihood that Madhava wrote a text on mathematics and astronomy from which quotations were given by later writers. However, this text is no longer extant.
17. The discussion that follows on Nilakantha's contribution is based on Ramasubramanian et al. (1994: 784–90) to which reference should be made for details.
18. According to one account of an unknown student, Nilakantha was born on 17 June 1444 and died in 1545.
19. The 21 pairs arise by taking, at a time, any two of the following seven quantities (a to g) as known from the right-hand side of the following equations:

$$x + y = a; x - y = b; xy = c; x^2 + y^2 = d; x^2 - y^2 = e; x^3 + y^3 = f; x^3 - y^3 = g$$

The solutions to 15 of the 21 pairs (${}^7\boldsymbol{C}_2$) are fairly straightforward while the remaining six are not. This throws light on some of the interesting mathematical contributions of this little-known mathematician.

20. The Variyars, as mentioned earlier, were a group of non-Brahmin temple officials who assisted the Brahmin priests in their religious rituals. A number of them were skilled in astrology and many were learned in Sanskrit. There is one story that they were descendants of a Brahmin and a Sudra woman.
21. Since the *Yuktibhasa* is the primary source referred to in elaborating on the mathematical content of this book, a separate appendix (Appendix I) is devoted to listing its contents in Chapter 6.
22. There are close similarities between this text and Sankara Variyar's (and Narayana's) *Kriyakramakari* and *Yuktidipika*, where the former is a *bhasya* on the *Lilavati* and the latter on the *Tantrasangraha.* However, Sankara Variyar acknowledges that the source of some material in the latter text is the work of Jyesthadeva (who he refers to as the Brahmin of Parakroda).
23. There were two common methods of telling time. At a point early in a child's education, the two methods were taught in the form of verses which were to be memorised. The *ativakyam* showed how to tell the time of the day by measuring the length of the shadow before and after noon. The *nakhstvakyam* showed how to reckon time at night by the position of stars and particularly by the time at which certain stars rose. This required considerable knowledge of astronomy and hence the method was only sketched out with further elaboration at an older age. At a later age, an *acharya* (teacher), usually a Pisarati, gave them further instruction on the use of water clocks for time reckoning where the basic unit of time was *narika* (24 minutes) or the time that a typical vessel took to sink.
24. In astronomical calculations, the longitude of a planet is measured along the ecliptic while, in fact, its motion takes place along its own orbit which deviates slightly from

the ecliptic. For an accurate computation of the planet's position, this deviation has to be corrected. Acyuta Pisarati gave the following formula (expressed here in modern notation) for the correction in the case of the moon:

$$R = \frac{\sin\alpha \cos\alpha(1-\cos\beta)}{\cos\theta}$$

where α is the longitudinal difference between the node and the planet, β the maximum latitude and θ the actual latitude.

25. His versatility is evident from the range of his compositions which included calculation manuals such as *Sphutanirnaya Tantra* (dealing with true computation of planets), the *Rasigolasphutaniti* (dealing with true longitude computation on the zodiac sphere), the *Chayastaka* (dealing with computation of gnomonic shadow), or his exposition of the *Drgganita* system in Karanottama or the computation of eclipses called the *Uparagakriyakrama* or *Uparagavimsati*—a commentary on the *Suryasiddhanta* or a commentary on Madhava's *Venvaroha.*
26. It is highly likely that this legend accompanied the Nambuthiris to Kerala for similar legends are associated with a number of other places with Brahmin settlements in Saurashtra, Maharashtra and Gujarat.
27. The women rarely travelled outside. In fact, the Malayalam word *antarjanam* for a Nambuthiri woman means 'woman inside' (the house).
28. A typical Nambuthiri male was dressed very simply with just a piece of cotton cloth around the waist, a towel around his shoulder and a thread around his torso signifying his Brahmin status. Marco Polo, during his travels in Kerala in 1288, was aghast to find bare-chested men of high status wandering around during the most formal of occasions.
29. In fact, the legend, as reported in *Keralopatti,* adds that the Nairs were ordered to be the 'eye' (*kan*), the 'hand' (*kei*) and at the 'command' (*kalpana*) of the Nambuthiris.
30. Nowhere else in India did such a complex and refined system of ritual pollution exist. A person from the lowest caste was not only untouchable but unapproachable in Kerala. That this form of rigid caste observances continued for a long time is borne out in the writing of the wife of a nineteenth-century English missionary, who gave the following touch and distance pollution parameters in descending order of caste hierarchy:

    > [A] Nair may approach a Brahmin but not touch a Brahmin; a Chogan Ezhava must remain thirty six paces off, and a Pulayen slave ninety six steps distant. A Chogan must remain twelve steps away from a Nair, and a Pulayan sixty six steps off, a Pariyan some distance further. A Syrian Christian may touch a Nair but they may not eat with each other. Pulayars and Pariyars, who are the lowest of all, may approach but not touch, much less may they eat with each other. (Hawksworth 1860: 8–9)

31. The date of the arrival of the Nambuthiris in Kerala is a matter of debate and has ranged from 700 BC to AD 1100. One of the widely quoted European commentators, William Logan (1906), argued that they arrived in Malabar in the early part of the

eighth century AD making their way through the Tulu country (in present-day Andhra Pradesh). This is supported by other sources who claim that it was Mayuravarman, the founder of the Kadamba dynasty, who invited Brahmins to settle in the Tulu-Kannada region. He ascended the throne around the middle of the fourth century.

32. The temples in Kerala can be broadly classified into two categories: those under the direct administration of the state and those managed by an *urnma*. The *urnma* consisted of heads of leading Nambuthiri households (or *illams*) living in areas around the temple. This form of private administration of the temple persists even today.
33. A study of the inscriptions on the copper plates of Tiruvalla, dated around the twelfth century, show the central temple as immensely wealthy and the pivot of the socio-economic life of the region. For a particular year, according to these inscriptions, the part of the temple land set aside for rice cultivation to 'feed the Brahmins' required 12,634 *kalams* (or 2.1 million litres) of rice seeds; and for the upkeep of 'eternal lamps' in the temple needed a seed capacity of over 2,000 *kalams* (or over 340,000 litres). Land was also allocated for financing a multiplicity of activities: daily food for the deity, purchase of ghee (or clarified butter), support of hospital and *sala* attached to the temple, conduct of festivals and ceremonies and payment to various grades of employees.
34. The *sabha* had another function which was no doubt highly remunerative: the power to fine a Nair 12 gold *kanams* (coins) for speaking ill of a Brahmin and 20 gold *kanams* for physical assault to a Brahmin.
35. In ancient Kerala, the *kalari* was a school where students were taught both martial arts as well as other conventional subjects. The teachers were generally Nairs and the system worked so long as the Nairs were in control. But when the Nambuthiris appeared on the scene with their superior Sanskrit learning, it was only a matter of time before the indigenous culture succumbed to its influence. Knowledge of Sanskrit became an essential requisite for any scholar. The Brahmins became the acknowledged teachers and the Nairs were forced to concentrate their attention on military training.
36. Only with the diversification of the communication media and increasing technological capability is questioning and self-expression encouraged and knowledge acquisition becomes informal and extensive.
37. *Gurukulas* developed in the families of aristocratic Brahmins, palaces of the chieftains and even in the houses of ordinary man who could afford to maintain it. They emerged between the *salas* of the early medieval period and the collegiate system of the modern period.
38. The *illams* of Kerala astronomers and mathematicians, such as Madhava, Paramesvara and Damodra, became important centres of learning.

3

The Mathematical Origins of the Kerala School

Introduction

Of all major mathematicians/astronomers of the 'classical' period (i.e., AD 450–1150) of Indian mathematics and astronomy, Aryabhata occupies a central place. His name is listed at the head of a group of individuals who made notable contributions to the development of Indian mathematics and astronomy. His impact is more substantial than his known contributions, for he founded a school that profoundly influenced the development of both astronomy and mathematics in Kerala and elsewhere for almost a thousand years.

There are a number of unresolved questions regarding him, including his place of birth and indeed his very identity. There is some confusion arising from the fact that more than one mathematician bearing the name of Aryabhata lived over a short period of time. There was one Aryabhata of Kusumpura who wrote the famous text *Aryabhatiya* in AD 499 and another Aryabhata who composed an astronomical text entitled *Maha Aryasiddhanta* around AD 950. Both the Indian Brahmagupta (b. 598) and the Persian al-Biruni (b. 973) mention a possible third Aryabhata who lived before the Aryabhata of Kusumpura. In his *History* written in 1036, al-Biruni refers to what we may now identify as the elder Aryabhata and the Aryabhata of Kusumpura. The existence of an elder Aryabhata may help to resolve some difficulties. For example, Brahmagupta has some harsh words in his

Brahmasphutasiddhanta (AD 628) for certain planetary theories of an Aryabhata. However, in an astronomical text he wrote 27 years later (*Khandakadyaka*), there are references in highly respectful terms to the work of an Aryabhata. Since the positions of planets and other heavenly bodies found in Brahmagupta's earlier text are very different from those that may be inferred from *Aryabhatiya*, one may conclude that the object of Brahmagupta's admiration may very well have been an older Aryabhata and the object of his denigration, the Aryabhata of Kusumpura who composed *Aryabhatiya*.

There is a tendency among historians of Indian mathematics to recognise the later two Aryabhatas and ignore the earliest one. The Aryabhata of Kusumpura is usually referred to as Aryabhata I and his year of birth is now accepted to be AD 476 since he mentions that he was aged 23 when he wrote *Aryabhatiya*. The second, a relatively minor figure known for only his commentary, *Maha Aryasiddhanta*, is usually referred to as Aryabhata II. We will henceforth refer to the author of *Aryabhatiya* as Aryabhata I or simply as Aryabhata.

The other unresolved question is regarding Aryabhata I's place of birth. This has become a matter of some contention because of the claims that he was born in Kerala. Kusumpura was not his place of birth but the place where he worked; there are many references, including one in Nilakantha's commentary on *Aryabhatiya*, composed around AD 1500, that he was born in a place called Asmaka. It is the location of these two places that has aroused controversy. One view is that while Kusumpura was a place in the north, probably near modern Patna, Asmaka could refer to a place in the south, probably Kerala. This view of the location of Aryabhata's birthplace has insufficient evidence to support it.

While there is controversy regarding his place of birth, there is sufficient evidence to indicate that Aryabhata spent a good part of his life in Kusumpura after attending the University of Nalanda. Kusumpura was one of the two major mathematical centres of ancient India, the other being Ujjain. The neighbouring Pataliputra, the capital of the Gupta empire at the time of Aryabhata, was the centre of a communications network that allowed information from other parts of the world to reach it easily, and also the mathematical and astronomical discoveries made by Aryabhata and his school to reach across India and also eventually into the Islamic world.

The Mathematics in *Aryabhatiya*

Irrespective of whether Aryabhata was a native of Kerala or not, there is little doubt that his text, *Aryabhatiya*, had a profound influence on the mathematicians/astronomers of Kerala and beyond. Within Kerala, it occupied a position akin to Euclid's *Elements* in Western mathematics or *Chiu Chang Suan Shu* ('Nine Chapters on the Mathematical Arts') in Chinese mathematics. An understanding of the content of *Aryabhatiya* is an essential prerequisite for a detailed examination of the work of the Kerala School in later chapters.

The *Aryabhatiya* deals with both mathematics and astronomy in 121 stanzas in all, written in an aphoristic style marked by brevity and conciseness.[1] There are four 'chapters' or sections (*padas*), of which the second concentrates on *ganita* or mathematics. The first chapter of the book consists of 13 stanzas in two parts setting out basic definitions and important astronomical parameters.

The next chapter, *Gitikapada* (or 'mathematical section'), contains 33 stanzas giving 66 rules and results in arithmetic, mensuration and geometry, trigonometry, algebra and mechanics without any proofs.[2] In the arithmetic section (Verses 1–5) appear the methods of finding the square and cube of any number,[3] method of extracting square and cube roots,[4] operations with fractions,[5] the Rule of Three and the method of inversion.[6] The rules contained in this section are not new; they were part of the tools available at that time, although Aryabhata expressed them in a concise but cryptic form that could be easily memorised.[7]

The section on geometry and mensuration (Verses 6–12) deals with the square, the cube, the triangle, the trapezium, the pyramid, the circle, the sphere and gnomonic shadow. It is in this section, widely commented on by historians of mathematics in the past, that some errors are found mainly arising from wrong generalisations on the basis of analogies. For example, Verse 6 reads:

> The product of the perpendicular (dropped from the vertex on the base) and half the base gives the measure of the area of a triangle.
>
> Half the product of that area (i.e., of the triangular base) and the height is the volume of a six-edged solid (pyramid).

The rule regarding the volume of a right pyramid is clearly based on a false analogy with the area of a triangle.[8] The thinking behind this inaccurate rule may have been as follows. Consider a large equilateral triangle divided into four equal equilateral triangles.[9] Fold up the peripheral equilateral triangles over the central one to form a pyramid. It is obvious that the area of this pyramid is equal to the area of the large equilateral triangle. Now, the base area of this pyramid is one-quarter the area of the large triangle. So the conclusion is that six such pyramids (which properly reassembled will constitute a rectangular solid) together will have volume equal to half the product of the area of the original triangle multiplied by the height of the pyramid.

The controversy regarding the accuracy of Indian geometry continues into the next verse. Verse 7 is cryptic almost to the point of incomprehension. Two main interpretations have been offered: it contains the rule for finding the volume of a sphere or it contains the rule for obtaining the area of the surface of a hemisphere.[10] Following are the two different interpretations, the first favoured by the Aryabhatan School, including the Kerala mathematician and astronomer Paramesvara, and the second of a more recent vintage of which Elfering (1977) is a good example:

(i) 'Half the circumference multiplied by half the diameter, that is the area of the circle. The volume of the sphere is the product of the area of the circle and its (area's) square root.'
(ii) 'Half the circumference multiplied by half the diameter, that is the area of the circle. This one (i.e., the circumference) multiplied by its defining base (i.e., the radius) is exactly the surface (area) of the hemisphere.'

Symbolically: Given $A = \frac{1}{4}Cd \equiv \frac{1}{2}Cr$ where A = area, C = circumference, d = diameter and r = radius

(i) $V = A.\sqrt{A} \equiv \pi.r^2.\sqrt{\pi.r^2} \equiv \sqrt{\pi^3}.r^3$
(ii) Area of the hemisphere $= Cr = 2\pi r^2$

(i) is clearly wrong and probably based on a false analogy with the volume of a cube which is the area of the base multiplied by its square root.
(ii) is, however, correct.

Verse 8 gives correctly the rule for calculating the lengths of the segments of the diagonals of a trapezium and the area of a trapezium. This could indicate that there was an early familiarity with the fundamental property of similar triangles, the property being perceived as an axiom more or less intuitively.[11] There was also an early interest in isosceles trapezium which was the shape of one of the altars (*smasana*) where *soma* sacrifices were carried out, as indicated in the *Sulbasutras* (c. 500–800 BC). Verse 8 concludes with a cryptic statement of a method for determining the area of any figure by decomposing it into trapeziums, a method similar to the one used until recently in field-surveying (triangulation).

Verse 9 states: '... the chord of the sixth part of the circumference is equal to the radius' (Shukla 1976).[12] This is the particular case of inscribing a regular polygon, namely a hexagon, within a circle and follows from the result that the side of the regular hexagon is equal to the semi-diameter.[13]

Verse 10 contains the following statement: 'Add 4 to 100, multiply by 8 and add 62,000; the result is approximately the circumference of a circle whose diameter is 20,000' (Shukla 1976).

This verse together with the previous one as well as the next one relating to the computation of a sine table are critical to the subsequent development of Indian mathematics, particularly by the Kerala School. The proximity of Verse 9 to Verse 10 may indicate that Aryabhata was aware of the common approach to 'squaring the circle', that is, finding a regular polygon of sufficiently large number of sides so that its perimeter is approximately equal to a circle of the same area. This approach begins with a polygon whose side was a 'chord of the sixth part of the circumference' (Shukla 1976). It may be deduced that Aryabhata used a regular inscribed polygon of 256 sides, to arrive at the accurate implicit value for π as $\frac{62832}{20000} = 3.1416$.[14]

Verse 11 explains the computation of the Indian Sine (RSine) table thus:[15] 'Divide a quadrant of the circumference of a circle (into as many parts as desired). Then from (right-handed) triangles and quadrilaterals, once can find as many RSines of equal arcs as one likes, for any given radius' (Shukla 1976).

Bhaskara I (fl. AD 600), the first of the great commentators of Aryabhata's work, gives three examples to illustrate the verse mentioned above. The first one concerns the computations of six RSines at intervals of 15 degrees in a circle of radius 3438 minutes.[16] The next involves the computation of 12 RSines at intervals of 7 degrees 30 minutes in a circle of radius 3438 minutes. The last example provided the prototype for later Indian Sine tables, the construction of 24 RSines at equal intervals of 3 degrees and 45 minutes in the circle of radius R (= 3438′). A discussion of the construction of these sine tables will be postponed to Chapter 5.

Verse 12 concerns the derivation of sine-differences:

> The first RSine divided by itself and then diminished by the quotient gives the second Rsine-difference. The same first RSine diminished by the quotients obtained by dividing each of the preceding R sines by the first RSine gives the remaining Rsine-differences. (Shukla 1976)

Expressed in modern notation, let $R_1, R_2, \ldots, R_{24}$ denote the 24 RSines and $\partial_1, \partial_2, \partial_3, \ldots, \partial_{24}$ denote the 24 sine-differences. The aforementioned rule (and especially the second sentence) has been interpreted in different ways.[17] One interpretation, favoured by the commentator Suryadeva (b. 1191) and others, gives

$$\partial_2 = R_1 - \frac{R_1}{R_1}$$

$$\partial_{n+1} = R_1 - \frac{R_1 + R_2 + \ldots + R_n}{R_1}$$

Other interpretations of the second sentence include those by (i) the Kerala mathematician, Paramesvara (fl. 1431), as reported in Singh (1939: 88); (ii) the Kerala mathematician Nilakantha (fl. 1500); and (iii) a commentator on the *Suryasiddhanta*, Raganatha (1603).

1. $\partial_{n+1} = \partial_n - \dfrac{\partial_1 + \partial_2 + \ldots + \partial_n}{R_1}$
2. $\partial_{n+1} = \partial_n - \dfrac{R_n}{R_1}(\partial_1 - \partial_2)$
3. $R_{n+1} = R_n + R_1 - \dfrac{R_1 + R_2 + \ldots + R_n}{R_1}$

The equivalence of (i) – (iii) is easily established. Nilakantha provides a geometrical demonstration of (ii) that is interesting in the light of later work on infinite series in Kerala. These interpretations, especially the one of Nilakantha, will be discussed in Chapter 5.

The next few verses deal with practical methods of constructing geometrical shapes such as circles, triangles and quadrilaterals as well as shadow problems.[18] Verse 17 begins with a statement of the Pythagorean theorem known in India at least from the time of the Baudhayana's *Sulbasutra* (c. 800 BC). This is then applied to the inner segments of a circle (see Figure 3.1), more particularly the result that in a circle of diameter AB and a chord CD with which the diameter intersect at right angles at E, $AE \cdot EB = (\frac{1}{2}CD)^2$. Bhaskara I, the commentator, has a well-known word problem involving a hawk and a rat to illustrate the use of this formula:

> A hawk, sitting on a wall of height 12 *hastas* sees a rat 24 *hastas* away at the foot of the wall. As the rat runs towards a hole in the wall where it lives, the hawk dives down and kills the rat. How far is the rat away from its hole when it is killed and what is the distance travelled by the hawk before the kill? (Keller 2006)

In Figure 3.1, AB represents roaming ground of the rat whose death occurs at the spot marking the centre of the circle (point O). The hypotenuse OC is the hawk's path. EB is the distance that the rat has to travel to reach the safety of its home. The square of the height of the

Figure 3.1

The Rat and Hawk: An Application of the Pythagorean Theorem

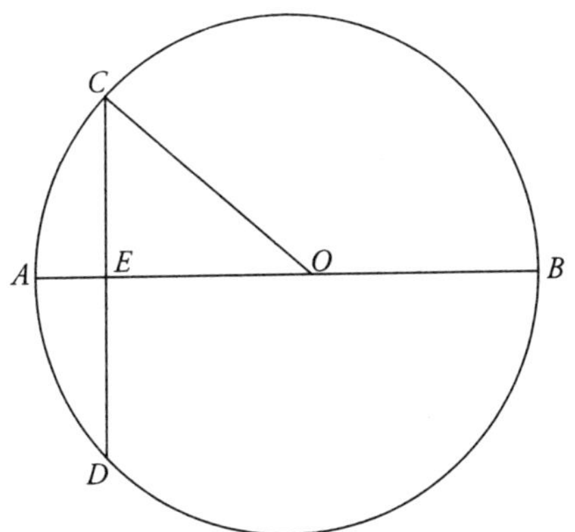

Source: Figure 45 of Keller (2006, Vol 1: p. 86).

hawk's position is $(\frac{1}{2}CD)^2 = 12^2 = 144$. The size of the rat's roaming ground is $AB = AE + EB = 24$. Estimate the quotient $144/24 = 6$. Add and then subtract this quotient from the roaming ground of the rat to get 30 and 18 *hastas* (1 *hasta* = 1.5 square foot) respectively. Their respective halves give the path of the hawk (OC) and the remaining distance that the rat has to cover to reach the safety of its hole.

Verse 18 relates to results regarding the arrows of intercepted arcs of intersecting circles and has little relevance for our purpose. Verse 19 states two ways of summing an arithmetical progression:

(i) 'The desired number of terms decreased by 1, halved, increased by the [number of terms] which precedes, multiplied by the common difference between the terms, plus the first term, is the middle term. This [result] multiplied by the number of terms desired is the sum of the desired number of terms.'

(ii) 'Or else, the sum of the first and last terms is multiplied by half the number of terms [is the result]' (Shukla 1976).

Symbolically, if n is the number of terms in an arithmetical progression extending from $(p + 1)$ to $(p + n)$ terms, d is the common difference, a and l are the first and last terms of the progression respectively, then we get the familiar formulae.

(i) $S = n[a + \{\frac{1}{2}(n - 1) + p\}d]$

(ii) $S = \frac{1}{2}(a + l)n$

Verse 20 contains a method of finding the number of terms (n) of an arithmetic progression:

> Multiply the sum of the progression by eight times the common difference, add the square of the difference between twice the first term and the common difference, take the square root of this, subtract twice the first term, divide by the common difference, add one, divide by two. The result will be the number of terms. (Shukla 1976)

Expressed in modern notation

$$n = \frac{1}{2}\left[\frac{\sqrt{8dS + (2a - d)^2} - 2a}{d} + 1\right]$$

For this formula to be known, there should have been a prior knowledge of the solution of quadratic equations of the form $ax^2 + bx + c = 0$. The same rules are given with slight variations by Bhaskara I (c. 600), Brahmagupta (c. 628), Mahavira (c. 850) and Sridhara (fl. 850–950).

Bhaskara I gives an example to illustrate the rule contained in Verse 19:

> The first term of a series is 2 and the common difference is stated to be 3. The [desired] number of terms is given as 5. Find the value of the middle term and the sum of all terms. (Shukla 1976)

The procedure suggested is to take the desired number of terms (5), decrease it by 1 (4), halve it (2), multiply by the common difference (6), increase this by the first term (8) which is the middle value. Just this multiplied by the desired number of terms produces the sum of this series as 40.

Note that this is equivalent to applying the formula $S = n[a + \{^1/_2 (n - 1) + p\}d]$ giving

$$\text{Sum} = 5[2 + \{^1/_2(5 - 1) + 0\}3] = 40$$

Verse 21 relates to the sum of a triangular series of the kind: $1 + (1 + 2) + (1 + 2 + 3) + \ldots$ to n terms. Symbolically

$$S = \frac{(n+1)^3 - (n+1)}{6}$$

Verse 22 contains an important result used later by the Kerala School:[19]

> The sixth part of the product of three quantities consisting of the number of terms, the number of terms plus one, and twice the sum of terms plus i is the sum of the squares. The square of the sum of the (original) series is the sum of the cubes. (Shukla 1976)

Symbolically:

$S_1 = \dfrac{n(n+1)}{2}$ where S_1 is the sum of the first n natural numbers

$S_2 = \frac{n(n+1)(2n+1)}{6}$ where S_2 is sum of squares of first n natural numbers

$S_3 = (S_1)^2 = \frac{n^2(n+1)^2}{4}$ where S_3 is sum of cubes of first n natural numbers

The last of the formulae, the sum of the cubes, may have been derived from the series

$$(3, 5), (7, 9, 11), (13, 15, 17, 19)\ldots$$

divided into the groups as shown by the brackets ( ). The sum of the numbers in each group is a perfect cube. Hence, the result follows easily by expressing the sum of n groups as the sum of natural odd numbers.

Both Brahmagupta and Bhaskara II give the aforementioned formula but Mahavira advances further by giving an expression for (*a*) the sum of the squares of the terms of an arithmetical progression; (*b*) the sum of the cubes of the terms of an arithmetic progression; and (*c*) the sum of series wherein each term represents the sum of a series of natural numbers up to a limiting number which is itself a member in a series in arithmetical progression, etc.

The work on series reported in Verses 19–22 in *Aryabhatiya* has a long history possibly going back to Vedic times, with an interest in the sum of a finite geometric progression found in the work of Pingala (c. 200 BC), in Jaina texts and the *Bakhshali Manuscript*. And the work would continue after Aryabhata in the work of his commentators and successors, notably Bhaskara I, the mathematician mentioned earlier, Narayana Pandit (fl. 1356) and finally the Kerala mathematicians, notably Nilakantha (c. 1443–1543).[20]

Verses 23–26 relate to the following rules of algebraic manipulations, including in (iii) the famous 'Rule of Three', a fundamental and well-used result in Indian mathematics.[21]

(i) 'Subtract the sum of the squares of two factors from the square of their sum. Half the result is the product of the two factors.'

(ii) 'Multiply the product (of two factors) by the square of two (or 4); add the square of differences between the two factors, take the square root, add and subtract the difference between the two factors, and divide the result by two. The result will be either of the two factors.'

(iii) 'In the Rule of Three (*trairasika*), multiply the *phala-rasi* (fruit) by the *iccha-rasi* (desire or requisition) and divide by the *pramana* (measure or argument). The required result *iccha-phala* (or fruit corresponding to desire or requisition) will be thus obtained' (Shukla 1976).

Symbolically, these rules can expressed thus, given a, b are the factors and p, i, m and f are *phala-rasi*, *iccha-rasi*, *pramana* and *iccha-phala* respectively.

(i) $ab = 0.5\left[(a+b)^2 - (a^2+b^2)\right]$

(ii) $0.5\left[\sqrt{4ab+(a-b)^2} \pm (a-b)\right]$ gives a or b

(iii) $f = {pi}/{m}$

It is likely that the identities defined in (i) and (ii) were intended to solve simultaneous equations of the types: $x + y = p$, $xy = q$; $x \pm y = p$, $x^2 + y^2 = q$; $xy = p$, $x^2 + y^2 = q$.

It is at this point that Aryabhata introduced a rule for finding the interest payable, given the principal (P), the amount (A) and the time (t). The rule is stated in *Aryabhatiya* thus:

> The interest on the capital, together with the interest [on the interest], with the time and capital for multiplier, increased by the square of half the capital. The square root of that, decreased by half the capital and divided by the time, is the interest on one's own capital. (Keller 2006)

Bhaskara I illustrates this rule with an example: 'The [monthly] interest on 100 is not known. However this interest increased by the interest [on the interest] for four months is 6. Find the monthly interest on a 100' (Keller 2006).

The procedure for solution is explained thus: The interest on capital increased by the interest (on the interest) amounting to a total of, 6 is multiplied by the time and the capital to obtain $6 \times 4 \times 100 = 2400$. Now square half the capital, 2500, and add to 2400 to obtain 4900. Find the square root of the result, 70, and decrease this by half the capital, 50, to get the result 20. Divide this result by the time, 4 months, to get the monthly interest on capital which is 5.[22]

Regarding the result (iii) given in the previous equation, this is the first time in Indian mathematics that the technical names for the 'Rule of Three' and for the four numerical quantities involved are given.[23] However, the succinct manner in which the rule is given would indicate that the 'Rule of Three' was already well known and that Aryabhata was merely restating it as a prelude to its use in astronomical computations. The antecedents of this rule has been traced back about a thousand years to a verse in the *Vedanga Jyotisa* (Kuppana Sastry 1985) and discussed in a paper by S.R. Sarma (2002: 135–36).

Bhaskara I's commentary on Aryabhata's work contains a detailed discussion of the rule and points to how the rule can be extended to encompass rules of five, seven, and so on. Bhaskara also introduces the question of the logical sequence in which the three numerical quantities should be set down and the order in which the multiplication (p times i) and division by m should be carried out. Later, Brahmagupta's formulation of the rule became a model for subsequent writers bringing out more explicitly the fact that the three quantities should be set down in such a way that the first and last be of like denomination and the middle one of a different denomination. This is reiterated by Sridhara (c. 750), Mahavira (c. 850) and Aryabhata II (c. 950) without adding much to the principles underlying the rule. However, Bhaskara II (1114–1185) in his *Lilavati* states the important point that nearly the entire arithmetic is based on the Rule of Three and that most of the topics dealt within *ganita* are but variations of this Rule of Three: 'Just as the universe is pervaded by Hari with His manifestations, even so all that has been taught [in arithmetic] is pervaded by the Rule of Three with its variations' (Apte 1937: 239). Again, in the *Siddhantasiromani*, Bhaskara II reiterates this point:

> Leaving squaring, square-root, cubing and cube-root, whatever is calculated is certainly a variation of the Rule of Three, nothing else. For increasing the comprehension of duller intellects like ours, what has been written in various ways by the learned sages … has become arithmetic. (Deva Sastri 1989)

And as we shall see in Chapter 6, the rule provided an essential tool in the derivation of infinite series by the Kerala mathematicians.

It is at this point that the *Aryabhatiya* discusses the rules of dividing one fraction by another and reducing all fractions to a common denominator. This serves as a prelude to his statement of the rule of inverse operations:

> In a reverse [operation], multipliers become divisors and divisors, multipliers, and additive [quantity] is a subtractive [quantity], a subtractive [quantity] an additive [quantity]. (Keller 2006)

To illustrate, consider an example from one of Aryabhata's commentators, Paramesvara:[24]

> What is the number multiplied by 3, divided by 5, the quotient increased by 6, the square root of the sum diminished by 1 and then squared yields 4? (Keller 2006)

The answer is obtained by reversing the order of the arithmetical operations to get

$$[(\sqrt{4} + 1)^2 - 6]5 \div 3 = 5$$

It was also at this juncture that Aryabhata introduces an important algebraic identity used particularly by the Kerala School. The statement is cryptic and difficult to express verbally. In modern notation

$$\frac{1}{n-1}\sum\left(\sum_1^n X_r^i - X_r\right) \equiv \sum_1^n X_r$$

The next subject dealt in *Aryabhatiya* appears to be one of the most important contributions of Aryabhata, namely his treatment of indeterminate equations. While there is some fragmentary evidence of earlier work on this subject, a systematic treatment of it only came with *Aryabhatiya*. The problem that Aryabhata addresses may be expressed in modern terms as follows:

> Find an integer (***N***) which when divided by another integer (***a***) leaves the remainder (r_1) and when divided by another integer (***b***) leaves the remainder (r_2).

And also the general problem:

> Find an integer (***N****) which being divided severally by the given integers $a_1, a_2, .. a_n$ leaves remainders $r_1, r_2, \ldots\ldots, r_n$ respectively

Symbolically, the two problems may be expressed thus

$$N = ax + r_1 = by + r_2 \quad (1)$$
$$N^* = a_1x_1 + r_1 = a_2x_2 + r_2 = \cdots = a_nx_n + r_n \quad (2)$$

If c denotes the difference between r_1 and r_2 in (1), then (1) may be rewritten as:

$$ax \pm c = by \quad (3)$$

It is suggested that c is always kept positive by appropriately labelling r_1 and r_2 such that $r_1 > r_2$.

A solution is offered in Verses 32 and 33. But there is some controversy as to how they are to be interpreted.[25] A discussion of this solution and its interpretations will take us far away from the scope of this book and will therefore not be considered here.

The *Ganitapada* ends with indeterminate equations. All subsequent sections of the *Aryabhatiya* relates to astronomy. The overall impression left by Aryabhata's mathematical section is that of a dry cryptic rendering of a collection of working rules necessary for solving a range of problems in practical life (such as surveying and commercial transactions) and in astronomy. The style is crisp without any trimmings. There is no attempt to entertain while teaching the rules that one finds both in works of Mahavira and Bhaskara II. The results are stated without any explanations. Yet *Aryabhatiya* marks a watershed in Indian mathematics and would remain the premier Indian text to be read and commented on for at least another thousand years. It would also become the fount of inspiration for the development of Kerala mathematics.

The Great Commentator: Bhaskara I

In our discussion of the contents of *Aryabhatiya*, it was pointed out that Bhaskara I provides useful illustrations and extension to what is often terse and difficult to understand. Little is known about Bhaskara I apart from his works which indicate that he was one of the more prominent members of the early Aryabhata School, whose three major works consisted of two astronomical treatises and a detailed commentary on *Aryabhatiya*. There is an inference by Shukla (1960) that since Bhaskara often refers to the *Asmakatantra* instead of the *Aryabhatiya* that he may have been a member of a school of mathematicians/astronomers situated in Asmaka, believed by some to be the birthplace of Aryabhata. There are other references to places in India in Bhaskara's writings. For example he mentions Valabhi (today's Vala in Gujarat), the capital of the Maitraka dynasty in the seventh century, and Sivarajapura, which were both in Saurashtra found today in the state of Gujarat on the west coast of the Indian subcontinent. Also mentioned are Bharuch (or Broach) in southern Gujarat and Thanesar in the eastern Punjab ruled by Harsha for 41 years from AD 606. This helps us to place Bhaskara in the early part of the seventh century, almost a century after Aryabhata.

As mentioned earlier, Bhaskara I was the author of two treatises, the *Mahabhaskariya* and the *Laghubhaskariya* and a commentary *Aryabhatiyabhasya.* The *Mahabhaskariya* is a work of eight chapters on Indian mathematical astronomy and includes topics which were fairly standard for such works at that time. It discusses topics such as the longitudes of the planets, conjunctions of the planets with each other and with bright stars, eclipses of the sun and the moon, risings and settings, and the lunar crescent. Chapter 7 of this text contains three verses that give the approximation formula for the sine function by means of a rational fraction which is amazingly accurate—a maximum error of less than 1 per cent. In other words, if we plot the actual sine values as well as Bhaskara's approximation to it, they are so close that we can't tell them apart! Indeed, rational approximations such as this are much better than power series in representing transcendental functions. Further discussion of this remarkable approximation formula that Bhaskara attributes to Aryabhata I is contained in the Appendix at the end of this chapter.

In AD 629, Bhaskara I wrote a commentary, the *Aryabhatiyabhasya*, on the *Aryabhatiya.* The commentary contains a discussion of only the 33 verses on mathematics in the original text. It provides a lucid elaboration of the solution of the indeterminate equations of the first degree and of the derivation of the trigonometric formulae and tables discussed earlier. In the course of his discussion, Bhaskara I considers how a particular rectangle can be treated as a cyclic quadrilateral. He was the first to discuss quadrilaterals with all the four sides unequal and none of the opposite sides parallel.

One of the approximations used for π for many centuries was $\sqrt{10}$. Bhaskara I criticised this approximation. He regretted that an exact measure of the circumference of a circle in terms of diameter was not available and he clearly believed that such a measure was not possible (or that π was irrational). Other subjects discussed in the *Aryabhatiyabhasya* include numbers and symbolism, the classification of different branches of mathematics, the names and solution methods of equations of the first degree, quadratic equations, cubic equations and equations with more than one unknown. Bhaskara I introduces the Euclidean algorithm method of solving linear indeterminate equations and the text contains detailed commentaries and examples illustrating Aryabhata's short aphoristic rules introduced earlier in this chapter. References are also made by Bhaskara I to the works of Indian mathematicians/astronomers earlier than Aryabhata.

The subsequent notable commentators on Aryabhata arose mainly in Kerala. Their work will be detailed in later chapters.

Appendix

Bhaskara I's Sine Approximation Formula

Bhaskara I's *Mahabhaskariya* is an eight-chapter text on Indian mathematical astronomy and includes topics such as the longitudes of the planets, conjunctions of the planets with each other and with bright stars, eclipses of the sun and the moon, risings and settings and the lunar crescent. Chapter 7 of that work contains three verses which give an approximation to the trigonometric sine function by means of a rational fraction. The rule is stated as follows:

> Without using the RSine-differences *makhi* (225), and so on, the rule is stated briefly. Subtract the degrees of the *bhuja* from the degrees of half a circle. Multiply the remainder by the degrees of the *bhuja* and set down the result in two places. Subtract the result in one place from 40500. By the fourth part of the remainder, divide the result in the other place multiplied by the radius. The *bahu* and *kotiphala* are thus obtained for Moon, Sun and other stars. Likewise the *krama* and *utkrama* (direct and versed sines). (Shukla 1960: vii, 17–19)

The formula is amazingly accurate leading to a maximum error of less than 1 per cent. In modern notation, the formula is

$$\text{Sin}x = \frac{4x(180-x)}{[40500-x(180-x)]} \tag{*}$$

where x is measured in degrees and $0 \leq x \leq 180$

Or measured in radians

$$\text{Sin}x = \frac{16x(\pi-x)}{[5\pi^2-4x(\pi-x)]}$$

Table 3.A.1

Applying Bhaskara I's Approximation Formula in Degrees

Degrees	Sinx	$\frac{4x(180-x)}{(40500-x(180-x))}$	Error
0	0	0	0
5	0.087155743	0.088328076	0.00117233
10	0.173648178	0.175257732	0.00160955
15	0.258819045	0.26035503	0.00153598
20	0.342020143	0.343163539	0.0011434
25	0.422618262	0.423208191	0.00058993
30	0.5	0.5	0
35	0.573576436	0.573041637	0.000534799
40	0.64278761	0.641833811	0.000953799
45	0.707106781	0.705882353	0.001224428
50	0.766044443	0.764705882	0.001338561
55	0.819152044	0.817843866	0.001308178
60	0.866025404	0.864864865	0.001160539
65	0.906307787	0.905374716	0.000933071
70	0.939692621	0.93902439	0.000668231
75	0.965925826	0.965517241	0.000408585
80	0.984807753	0.984615385	0.000192368
85	0.996194698	0.99614495	4.97482E-05
90	1	1	0

Source: Author.

Bhaskara attributes this formula to Aryabhata although it is not found in *Aryabhatiya* or his other extant works.

Table 3.A.1 gives the values of the Bhaskara's approximation and that of the values from an electronic calculator for x in the range $0 \leq x \leq 90$. Bhaskara's equation is simple, elegant and, moreover, gives values of the sine function accurate for practical purposes.[26]

How did Bhaskara I come up with this function? The start of the story may lie in the fact that the sine curve had roots $x = 0$ and $x = 180$ and its curve 'looks' like a quadratic function $f(x) = x(180 - x)$. However, $\sin 90° = 1$ while $f(90) = 8100$ so there must have been a scaling down of the function. And this scaling factor is not uniform because $\sin 30° = \frac{1}{2}$ while $f(30) = 4500$ which implies a scaling down factor of 9000 while the corresponding values for $x = 90$ imply a scaling down factor of 8100. In addition, $\sin 60° = \frac{1}{2}\sqrt{3}$ and $f(60) = 7200$ imply a scaling factor of $\frac{14400}{\sqrt{3}}$. Thus the scaling function is not linear as the difference scaling-down factors is –900 between $x = 30$ and $x = 90$, an average 'gradient' of $-900/60 = -15$, while the difference scaling-down factors is approximately -686 between $x = 60$ and $x = 90$, an average 'gradient' of $-686/30 = -22.87$.

Thus Bhaskara must have conjectured that the scaling-down function was quadratic. Let

$$\text{Sin}x = \frac{x(180-x)}{ax^2+bx+c} \qquad (^{**})$$

Now Indian mathematicians since the time of Aryabhata had known the values of the sines of a large number of angles. In particular they knew that Sine 90° = 1, Sine 30° = Sin 150° = ½, Sine 60° = ½√3. Substituting $x = 30°$ and $x = 150°$ in the above expression (**) gives

$$\frac{1}{2} = \frac{4500}{900a+30b+c} = \frac{4500}{22500a+150b+c}$$

Equating the denominators gives $21600a = -120b$ or $b = -180a$. Substituting $b = -180a$ and $x = 30$ in the expression (**) we obtain

$$1 = \frac{8100}{8100a-16200a+c}$$

Thus $c = 8100a + 8100$. With this value for c and with $x = 60$ we get[27]

$$\frac{\sqrt{3}}{2} = 0.866 = \frac{7200}{3600a-10800a+8100a+8000} = \frac{7200}{8100-900a} = \frac{8}{9-a}$$

This means that

$$a = \frac{8}{0.866} - 9 = 0.238 = 0.238 \text{ correct to 3 decimal places}$$

So Bhaskara I must have used the approximation $a = ¼$. Using this approximation and simplifying gives

$$f(x) = \frac{x(180-x)}{\left(\frac{1}{4}x^2 - \frac{180x}{4}\right) + \frac{8100}{4} + 8100} = \frac{4x(180-x)}{40500-x(180-x)}$$

This is identical to the approximation formula given in (*).

However, following the work of Aryabhata, Indian mathematicians calculated sine values for any angle in radians. It will now be shown that Bhaskara I's approximation formula (*) may be expressed in radians as

$$\sin\frac{\pi}{n}=\frac{16\,(n-1)}{5n^2-4n+4} \qquad (**)$$

To see how the formula (**) may have been derived we write, using the same reasoning as before, $\sin\frac{\pi}{n}$ as the ratio of two quadratics

$\dfrac{a_1+a_2n+a_3n^2}{b_1+b_2n+b_3n^2}$ where a_i and b_j for $i,j=1,2,3$ are constants (***)

The following properties of $\sin\frac{\pi}{n}$ holds when $n\geq 1$

(1) $\sin\frac{\pi}{n}=\sin\frac{\pi}{m}$ where $m=\dfrac{n}{n-1}$

(2) $\sin\frac{\pi}{n}\rightarrow 0$ as $n\rightarrow\infty$

(3) $\sin\frac{\pi}{n}=0, 1, 0.5$ when $n=1, 2, 6$ respectively

Using (2) we deduce that $a_3=0$
Then using (1) we get

$$(4)\quad \frac{(n-1)\{n(a_1+a_2)-a_1\}}{b_1-(b_2+2b_1)n+n^2(b_1+b_2+b_3)}\equiv\frac{a_1+a_2n+a_3n^2}{b_1+b_2n+b_3n^2}$$

Finally using (3) we derive the following relations:

(5) $a_1+a_2=0;\ b_1+2b_2+4b_3=a_1+2a_2;$
$b_1+6b_2+36b_3=2a_1+12a_2$

Substituting $a_1+a_2=0$ into (4) yields $b_1+b_2=0$. Whilst substituting the last two equations in (5) and using the derived relations $a_1=-a_2$ and $b_1=-b_2$ gives

$$4b_3-b_1=a_2$$
$$36b_3-5b_1=10a_2$$

Solving these equations we obtain

$$b_3=\frac{5}{16}a_2 \text{ and } b_3=\frac{5}{4}b_1$$

Substituting these relations between a_i and b_j for $i, j = 1, 2, 3$ into (***) gives

$$\frac{\frac{16}{5}(n-1)}{n^2 - \frac{4}{5}n + \frac{4}{5}} \equiv \frac{16\,(n-1)}{5n^2 - 4n + 4}$$ which is the approximation formula (**)

An interesting variant of (**) is given by the sixteenth-century Indian mathematician Ganesha

$$\sin\frac{\pi}{n} \approx \frac{n^2 - (n-2)^2}{n^2 + \frac{1}{4}(n-2)^2}$$

Making n the subject of the formula gives a procedure to calculate an angle for a given sine for all values where $n \geq 2$

$$1 - \frac{2}{n} = \sqrt{\frac{1 - \sin\frac{\pi}{n}}{1 + \frac{1}{4}\sin\frac{\pi}{n}}}$$

It is interesting that Bhaskara I's approximation formula reappears in a different guise in the work of his more famous namesake, Bhaskara II (b. 1114), who gives the formula for the approximate length of the chord of a circle (y) in terms of its diameter (d), arc length (a) and circumference (C). Expressed in modern notation, the formula is

$$y = \frac{4ad(C-a)}{1.25C^2 - a(C-a)}$$

Note that

1. $y = 0$ if $C - a = 0$ or $a = 0$
2. Since $y \propto a(C-a)$ and $y \propto d$, it would follow that $y \propto da(C-a)$
3. For a dimension formula in y, we need to divide by a quadratic in C and a

Define α, β, γ, δ such that

$$y = \frac{\alpha da(C-a)}{\beta C^2 + \gamma Ca + \delta a^2}$$

Table 3.A.2

Applying Bhaskara II's Approximation Formula in Radians

n	True Values (y)	Values from Bhaskara II's Formula (y^*)	Error ($y - y^*$)
1	1.414	1.412	0.002
2	0.767	0.765	0.002
3	0.518	0.521	−0.003
4	0.390	0.393	−0.003
5	0.313	0.316	−0.003
6	0.261	0.264	−0.003
7	0.224	0.227	−0.003
8	0.196	0.198	−0.002
9	0.174	0.177	−0.003

Source: Author.

To compute α, β, γ, δ, consider the following special cases:

1. $y = d$ when $a = C/2$
2. $y = d/2$ when $a = C/6$
3. $y = \sqrt{2}\, d/2$ when $a = C/4$
4. $y = \sqrt{3}\, d/2$ when $a = C/3$

To apply Bhaskara II's formula, take the values $\sqrt{2} = 24/17$ and $\sqrt{3} = 64/37$.

Table 3.A.2 contains a comparison between the actual values of y (chord lengths) and those computed using the approximation formula for $a = \dfrac{C}{4n}$ for $n = 1, 2, \ldots, 9$ and $d = 2$

The extraordinarily accurate approximations for the sine values expressed as radians is confirmed by the last column of Table 3.A.2 which gives the calculated error values.

Notes

1. At the time of Aryabhata, mathematics was rarely treated outside its astronomical context. One of Aryabhata's contributions was to have a separate section (or chapter) in his *Aryabhatiya* on mathematics entitled *Ganitapada*.

2. The concluding two sections after the section on mathematics consist of 25 verses on the reckoning of time and planetary models and 50 verses on the sphere and eclipses.
3. Both a geometrical representation and an arithmetical operation are implied in the terms 'square and squaring' and 'cube and cubing'. Thus Verses 3(a–b) and 3(c–d) may be translated respectively as: 'An equilateral quadrilateral with equal diagonals is a "square" (*samacaturasra*). The product of two equal quantities is also called "square" (*varga*).' 'The continued product of three equals and a (rectangular) solid having twelve (equal) edges is called a "cube"' (Shukla 1976).
4. The actual algorithms for calculating square and cube roots are explained in Verses 4 and 5 respectively. The methods are similar to the ones taught in schools in the period before the advent of electronic calculators. For details of these verses see Shukla's critical edition and translation of *Aryabhatiya* (1976: 36–38).
5. Verse 27(a–b) dealing with the simplification of the quotient of fractions ('the numerators and denominators of the multipliers and divisors should be multiplied by one another'), follows from the discussion of the 'Rule of Three' in the previous verse. Verse 27(c–d) discusses the rule for the reduction of two fractions to a common denominator: 'Multiply the numerator as also the denominator of each fraction by the denominator of the other fraction').
6. The important 'Rule of Three' is stated clearly in Verse 26; and the method of inversion is stated succinctly in Verse 28: 'In the method inversion multipliers become divisors and divisors become multipliers, additions become subtraction and subtraction becomes addition.'
7. *Patiganita* was the name given to that branch of Indian mathematics which dealt with arithmetic and mensuration. The subject seems to have attained an independent status during the early centuries of the Common Era and combined with geometry it became an important part of the practical arts and crafts such as building houses, estimating volume of piles of sawn timber and of bricks and stones, or estimating amounts of grains contained in piles, and so on. Geometry, from as early as the recording of the *Sulbasutras* (500–800 BC), was never a distinct subject but always went in hand in hand with algebraic formulations and computational techniques.
8. The correct rule is found in Verse 44 of the *Brahmasphutasiddhanta* of Brahmagupta (b. 598). It states: 'The volume of a uniform excavation divided by three is the volume of the needle-shaped solid.'

   In modern terminology, this is equivalent to: volume of a pyramid = 1/3(area of base) × height.
9. The reader is invited to follow these instructions with the help of a folded paper and identify the 'concrete' thinking and the fallacy behind this derivation.
10. This problem bears some resemblance to Problem 10 from the Moscow Papyrus from ancient Egypt (c. 1850 BC). It is interesting that one of the interpretations of the Egyptian problem relates to finding the surface area of a hemisphere given the diameter. For further details of this Egyptian problem, see Joseph (2000: 88).
11. In early Indian geometry, there is no theory of parallels but there is sufficient evidence to show that there existed, instead, a theory of similar triangles. It is more than likely that there had developed an early intuitive perception of the validity of the postulate

that two intersecting straight lines cutting two parallel straight lines form triangles whose corresponding sides are proportional. This axiom then became the basis of much of Indian geometry, especially the well-known property of the right-angled triangle attributed to Pythagoras.

12. In modern notation, this means chord $60° =$ Radius (R) or $R\sin 30° = R/2$. Nilakantha provides an interesting geometrical demonstration (see Chapter 4).
13. This is easily deduced from the fact that the regular (equal-sided) hexagon is made up of six equilateral triangles where the side of the hexagon is equal to the radius of the circle. A geometrical demonstration of this result is contained in *Aryabhatiyabhasya* of Nilakantha and shown in page 70 of next chapter.
14. The method involves approximating the circle to a succession of regular polygons having increasing number of sides. This involves finding length of sides of regular polygons of '$2n$' sides from a polygon of 'n' sides. We start initially by calculating the length of the side of an octagon from a square and then proceed to the length of a side of a 16-sided polygon from octagon, 32-sided polygon from 16-sided polygon and so on. When the number of sides (n) becomes very large, the polygon will approximate to circle. At the beginning of the chapter on circle in the *Yuktibhasa* (c. 1530), this method is referred to as '*square –square root method*' since it essentially makes use of the relation between the sides of a right-angled triangle.

    The implict value for π given in *Aryabhatiya* of 62832/20000 is arithmetically equivalent to the value given in Bhaskara II's *Lilavati* (AD 1156) as 3927/1250. Now the side of a 32-sided regular polygon that can be inscribed in a circle having diameter 1250 units will be 123.2386. The side of a 64-sided regular polygon that can be inscribed in a circle having diameter 1250 units will be 61.4085. The side of a 128-sided regular polygon that can be inscribed in a circle having diameter 1250 units will be 30.6857. The side of a 256-sided regular polygon that can be inscribed in a circle having diameter 1250 units will be 15.3405. Assuming the circumference of the circle and the perimeter of the regular polygon having 256 sides are approximately equal, the circumference is the product of 256 and 15.3405 = 3927.168. Then value of π is 3927 /1250 =3.1416. The impressive earlier work of the Chinese mathematicians, Liu Hui (c. AD 260) and Tsu Chung Chih (c. AD 480), along similar lines should be recognised. For further details, see Joseph (2000: 193–96).
15. The Indian 'Rsine' in this context refers to the product of radius (R) and sine. What is interesting here is that this is the first recorded instance of the Indian use of the semi-chord rather than the whole chord which was a characteristic of the Greeks. Despite the seminal work of Ptolemy in the development of trigonometry, he failed to appreciate the innovative nature of the semi-chord-based sine function which arose in India a few centuries later.
16. This value is easily explained with present-day notation. For a given circle of circumference 360 degrees, the radius of the circle is $360/2\pi = 57.2957$, where π was approximated by the Aryabhatan value of 3.1416 implied in the previous verse. Now, if the circumference was measured in minutes (60 minutes = 1 degree), then the total circumference is 21600 minutes and the corresponding radius is 3437.7467... or approximately 3438. As we will see, this value became the standard measure for the radius of a circle in the construction of sine tables.

17. This rule is equivalent to the modern well-known differential formula

$$\frac{d^2(\text{Sin}x)}{dx^2} = -\text{Sin}x$$ where 'Sine' is equivalent to the modern sine multiplied by the radius 3438.

According to this rule, each sine-difference diminished by the quotients of all previous differences and itself by the first difference (i.e., 225) gives the first difference. The differences given in the text are: 225, 224, 222, 219, 215, 210, 205, 199, 191, 183, 174, 164, 154, 143, 131, 119, 106, 93, 79, 65, 51, 37, 22, 7. The same results are also given in the *Suryasiddhanta* applying the same rule for obtaining them.

18. The topic examined here are the mathematics of the sun-dial and shadows. After a preliminary examination of rules for constructing a circle and a triangle (or more likely an equilateral triangle) given a side, and a rectangle or square given the diagonal, instructions are offered for determining experimentally the horizontal and vertical planes by means of water and the plumb-line respectively. For determining the radius of the gnomon-circle given the height of the gnomon and the length of the shadow, the Pythagorean rule is recommended. The main discussion relates to the two rules for determining the lengths of the shadow of a gnomon of given height, and for determining the height of the source of light and its distance from two equal gnomons casting given length of the shadows.
19. The same rules are repeated in the seminal texts of Kerala mathematics, notably Nilakantha's *Aryabhatiyabhasya*, Jyesthadeva's *Yuktibhasa* and Sankara and Narayana's *Kriyakramakari*. They will make their reappearances in Chapters 4 and 6 of this book.
20. For further discussion of Indian work on series, see Saraswati Amma (1963) and Singh (1936).
21. For example, its use in the estimation of the circumference of a circle in the *Yuktibhasa* is discussed in what is conventionally labelled as Chapter 6 (Sarma 2008: 45–82). It must be remembered that the original text is one whole work not divided into any separate sections. The classification into different sections (or chapters) is based on the topics covered. For further details, see Appendix I of Chapter 6 of this book.
22. A verification of this result is shown using a 'Rule of 5'. In modern terms, the rule is obtained from solving the quadratic equation

$$Pr^2t^2 + Prt = A \text{ where } r \text{ is the interest rate}$$

23. The relevant passage reads: 'The known result is to be multiplied by the quantity for which the result is wanted, and divided by the quantity for which the known result is given' (Shukla 1976). This concise instruction may be expressed in terms of the terms introduced by Aryabhata as: Take the known result *phala-rasi*, multiply it with *iccha-rasi* and divide it by *pramana* to get the result to be known as *iccha-phala*.
24. A better known problem whose solution involves the 'method of inversion' is wrongly attributed to Aryabhata although it appears in Bhaskara II's (b. AD 1114) *Lilavati*. The reader may wish to try and solve it:

> Oh beautiful maiden with beaming eyes, tell me, since you understand the method of inversion, what number multiplied by 3, then increased by

three-quarters of the product, then divided by 7, then diminished by one-third of the result, then multiplied by itself, then diminished by 52, whose square root is then extracted before 8 is added and then divided by 10, gives the final result of 2? (Apte 1937)

25. The interpretations include those of Rodet (1879), Kaye (1908), Heath (1910), Majumdar (1912), Sengupta (1927), Ganguly (1928), Clark (1930) and Datta (1932). The translations of Rodet and Kaye are now accepted as faulty. But the damage persisted with the adoption of Kaye's interpretation by certain Western and Indian historians of mathematics, namely, in this particular case, Heath and Majumdar. Sengupta's interpretation is based on Brahmagupta's; Clark's on that of Paramesvara's; and Datta's and Ganguly's which refer to Bhaskara I's are now acknowledged to be the more satisfactory ones.
26. It may be argued that this is the best rational approximation that can be devised.
27. Note, that ever since Aryabhata devised a method to calculate square roots, Indian mathematicians could approximate sine 60 by a rational number, that is, sine 60 = 0.866.

4

The Highlights of Kerala Mathematics and Astronomy

Until recently there was a misconception that mathematics in India made no progress after Bhaskara II who lived in the twelfth century AD and that later scholars seemed 'content to chew the cud', writing endless commentaries on the works of venerated mathematicians who preceded them, until they were introduced to modern mathematics by the British. Though the picture about the rest of India is somewhat patchy, in Kerala, the period between the fourteenth and seventeenth centuries marked a high point in the indigenous development of astronomy and mathematics. The quality of the mathematics available from the texts that have been studied is of such a high level compared with the earlier period that it is difficult to bridge the gap between the two periods. Nor can one invoke a 'convenient' external agency, like Greece or Europe to explain the Kerala phenomenon. These were later discoveries in European mathematics, which were anticipated by Kerala astronomer-mathematicians 200 to 300 years earlier.

In 1834, Charles Whish published an article in which he referred to four works—Nilakantha's *Tantrasangraha*, Jyesthadeva's *Yuktibhasa*, Putumana Somayaji's *Karanapaddhati* and Sankara Varman's *Sadratnamala*—as being the key astronomical and mathematical texts of the Kerala School. While there were some doubts about Whish's views on the dating and authorship of these works, his main conclusions are still broadly valid. Writing about *Tantrasangraha*, he claimed that this work laid the foundation for a complete system of 'fluxions'.[1] Referring to the *Sadratnamala*, Whish claims that it 'abounds in fluxional forms and series to be found in no work of foreign countries'. The Kerala discoveries highlighted by Whish included the Gregory series for the inverse tangent, the Leibniz power series for π, and the Newton power series for the sine and cosine, as well as certain remarkable rational approximations of trigonometric functions, including the well-known Taylor series approximations for the sine and cosine functions. And these results had apparently been obtained without the use of infinitesimal calculus.

In the 1940s it was Rajagopal and his collaborators who highlighted the contributions of Kerala mathematics, although few of their results percolated into the standard Western histories of mathematics. For example, Boyer (1949: 244) wrote that 'Bhaskara (i.e., Bhaskara II) was the last significant medieval mathematician from India, and his work represents the culmination of earlier Hindu contributions'. And according to Eves (1983: 164), 'Hindu mathematics after Bhaskara made only spotty progress until modem times.'

Madhava's pioneering work on power series for π and for sine and cosine functions is referred to by a number of the later writers of the Kerala School, although the original sources remain undiscovered. Nilakantha was mainly an astronomer, but his *Aryabhatiyabhasya* and *Tantrasangraha* contain work on infinite-series expansions, problems of algebra and spherical geometry. Jyesthadeva produced his seminal work, *Yuktibhasa*, one of those rare texts in Indian mathematics and astronomy that gives detailed demonstrations of many theorems and formulae in use at the time.[2] This work is mainly based on the *Tantrasangraha* of Nilakantha. A joint commentary on Bhaskara II's *Lilavati* by Sankara Variyar and Narayana (fl. 1500–1575), entitled *Kriyakramakari*, also contains a discussion of Madhava's work. The *Karanapaddhati* by Putumana Somayaji (fl. 1660–1740) provides a discussion of the various trigonometric series. Finally there is Sankara Varman, the author of *Sadratnamala*, who lived at the beginning of the nineteenth century and may be said to have been the last of the notable names in Kerala mathematics. His work in five chapters contains, appropriately, a summary of most of the results of the Kerala School, without any proofs though.

Astronomy provided the main motive for the study of infinite-series expansions and rational approximations of circular and trigonometric functions. For astronomical work, it was necessary to have both an accurate estimate of π and highly detailed trigonometric tables. In this and other areas the members of the Kerala School made some significant discoveries.

A list of the principal achievements of the Kerala School in the wider context of the general history of mathematics and astronomy would include:

1. *The first correct formulation of the equation of the centre* for the interior planets, Mercury and Venus, by Nilakantha (fl. 1444–1545) in the *Tantarsangraha* about one hundred years before the German astronomer, Johannes Kepler (1571–1630).[3]
2. *The discovery of the formula for the 'reduction to the ecliptic'*,[4] first given in the West by Tycho Brahe (1546–1501), but in Kerala it was discovered by his contemporary, Acyuta Pisarati (1550–1621), in his book *Sphutanirnaya.*
3. Nilakantha's commentary on *Aryabhatiya* contains a number of interesting geometrical demonstrations. These include showing that:

   (i) *The area of a circle is equal to the product of half its circumference and half the diameter.*

This is done by cutting up a circle into a large number of equal tapering laminas (see Figure 4.1(a)) and on that basis the base of each lamina (being a small arc segment) will approximate a straight line; juxtaposing each pair of two thin laminas (see Figure 4.1(b)) to form a series of rectangles

Figure 4.1

Area= ½ Circumference × ½ Diameter

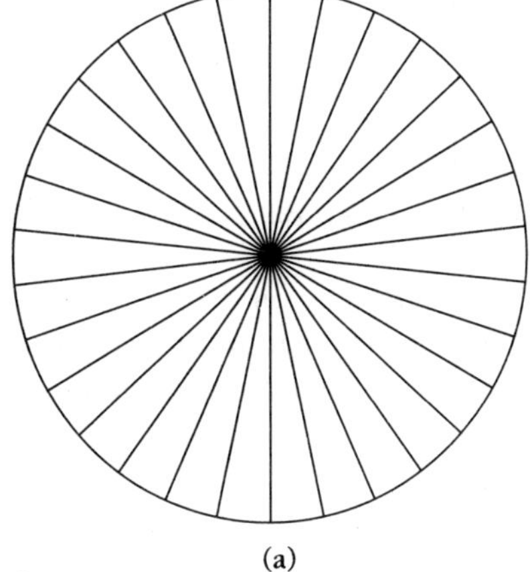

(a)

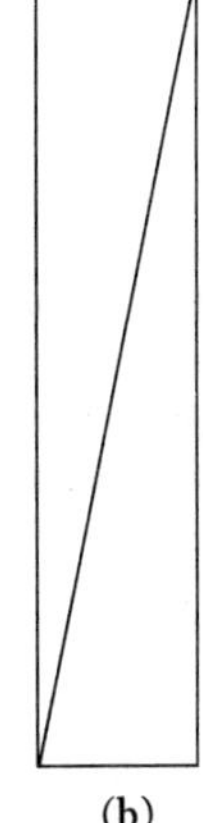

(b)

with the longer side being equal to the radius and the smaller being the arc segment. These rectangles are then re-arranged into a rectangular sheet. From this one can deduce that the area of the circle is equal to the area of the rectangular sheet. Note that the area of the rectangular sheet is the product of the adjacent sides consisting of the product of half the circumference and of the radius (or half the diameter).

(ii) *The chord of one-sixth of the circumference of a circle is half the diameter.*

The demonstration of this result follows directly from the following diagram (see Figure 4.2) containing three equilateral triangles whose side is a chord of length radius.

Figure 4.2

The Chord of a Circle

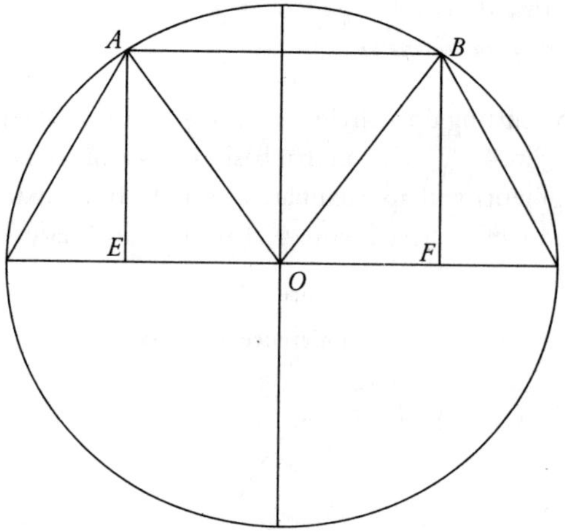

A
B
E
F
O

(iii) *A formula for the sum of an infinite convergent geometric series*

First given by Nilakantha in his commentary on the *Aryabhatiya*, this rule was used in deriving an approximation for an arc of a circle in terms of its chord. The rule has been stated as follows (Saraswati Amma 1963: 325–26):

> The sum of an infinite series, whose terms (from the second onwards) are obtained by diminishing the preceding ones by the same divisor (i.e., by

the denominator of the first term) is always equal to the numerator of the first term divided by one less the common divisor.

Or expressed in modern notation, this is equivalent to the modern rule for the summation of a convergent infinite geometric series. That is, if the first term is a and r is the common ratio (or divisor), the geometric series generated is of the form

$$a + ar + ar^2 + ar^3 + \dots$$

The sum of the series, assuming that it is convergent (i.e., $-1 < r < 1$), is

$$S = \frac{a}{1-r}$$

To illustrate the application of the Nilakantha Rule, consider the following infinite convergent geometric series:

$$\frac{1}{4} + \frac{1}{4^2} + \frac{1}{4^3} + \dots$$

What is to be demonstrated is that

$$\frac{1}{4} + \frac{1}{4^2} + \frac{1}{4^3} + \dots = \frac{1}{3}$$

The sequence of steps in the demonstration is as follows:

$$\frac{1}{3} = \frac{1}{4} + \frac{1}{4 \times 3}$$

$$\frac{1}{4 \times 3} = \frac{1}{4 \times 4} + \frac{1}{4 \times 4 \times 3}$$

$$\frac{1}{4 \times 4 \times 3} = \frac{1}{4 \times 4 \times 4} + \frac{1}{4 \times 4 \times 4 \times 3}$$

The summation of this series may be expressed as

$$\frac{1}{4} + \frac{1}{4^2} + \frac{1}{4^3} + \dots = \frac{1}{3} - \left[\frac{1}{4 \times 3} + \frac{1}{4 \times 4 \times 3} + \frac{1}{4 \times 4 \times 4 \times 3} + \dots \right]$$

$$= \frac{1}{3} - \left[\sum_{n=1}^{\infty} \frac{1}{(3)(4)^n} \right]$$

If n is very large, the terms in the square bracket will become negligibly small and can be ignored.[5] Therefore, the sum of this infinite series equals $\frac{1}{3}$ when $n \to \infty$. This rule was applied, as we will see later, in the derivation of the arctan (and π) series in Chapter 6.

(iv) *Nilakantha's geometrical approach to the summation of an arithmetical series*

Each term of the arithmetical series is represented by a rectangular strip (see Figure 4.3(a)) whose length is equal to the number itself and whose width is one unit. Each of the strips is arranged in a manner given in Figure 4.3(a) such that the piling up of the rectangles represents the series and the area of the figure is the sum of the series. Now assume that, as shown in Figure 4.3(b), the piling up of the rectangular strips are fitted together, with one inverted to allow for such a fit. Now the adjacent sides of the whole rectangle in Figure 4.3(b) are given by the number of rectangular strips and the sum of the first (a) and the last term (t) of the series. So the area of the whole rectangle is $n(a + t)$. So the sum of the arithmetic series is $\frac{1}{2}n(a + t)$.

Nilakantha uses a similar form of geometrical reasoning in his *Aryabhatiyabhasya* to obtain the sum of a series of triangular numbers,[6] the sum of sums,[7] and so on. This technique was then extended by Sankara and Narayana in their *Kriyakramakari* and in Jyesthadeva's *Yuktibhasa*. For further details, see Mallayya (2002).

Figure 4.3

Summation of Arithmetical Series

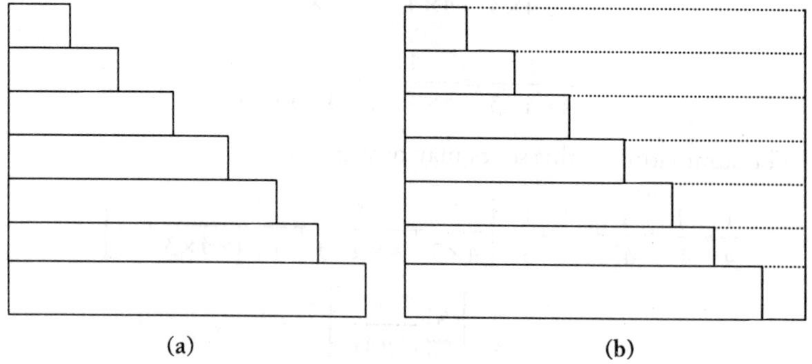

4. *The discovery of Newton–Gauss interpolation formula to the second order* by Govindasvamin (c. 800–850) at a date earlier than that of either of the names associated with the formula. This formula may be expressed in modern notation as

$$f(x+h) = f(x) + n\Delta h f(x) + \frac{n(n+1)}{2}[\Delta f(x) - \Delta f(x-h)]$$

where Δ is the finite difference operator and h is the step-length.

A consistent strand running through the work of Indian mathematicians, at least from the time of Aryabhata I (b. AD 476), was the search for more accurate methods of interpolation, particularly in respect of sine values for intermediate angles. In fact, Brahmagupta (b. AD 598) used a second difference interpolation formula which was rediscovered nearly a thousand years later and labelled as Newton–Stirling formula.

5. In Paramesvara's commentary on Bhaskara II's *Lilavati* occur a number of formulae relating to a cyclic quadrilateral and in particular for *obtaining the circum-radius of a cyclic quadrilateral whose sides are known.*[8] The cyclic quadrilateral was an important device used by the Kerala School for deriving trigonometric results, including

$$\sin^2 A - \sin^2 B = \sin(A+B)\sin(A-B)$$

and

$$\sin A \,.\, \sin B = \sin^2 \tfrac{1}{2}(A+B) - \sin^2 \tfrac{1}{2}(A-B)$$

Paramesvara states the formula thus:

> The three sums of the products of the sides, taken two at a time, are to be multiplied together and divided by the product of the sums of the sides taken three at a time and diminished by the fourth. If a circle is drawn with the square root of this quantity as radius the whole quadrilateral will be situated in it. (Sarasvati Amma 1979: 108)

Expressed in modern notation, if a, b, c, d are the sides of a cyclic quadrilateral shown in Figure 4.4, and r is its circum-radius with circle centre at O and $x = AC$ and $y = BD$ are its diagonals, then

$$x = \sqrt{\frac{(ac+bd)(ad+bc)}{(ab+cd)}}; \quad y = \sqrt{\frac{(ab+cd)(ac+bd)}{(ad+bc)}}$$

Figure 4.4

Circum-radius of a Cyclical Quadrilateral

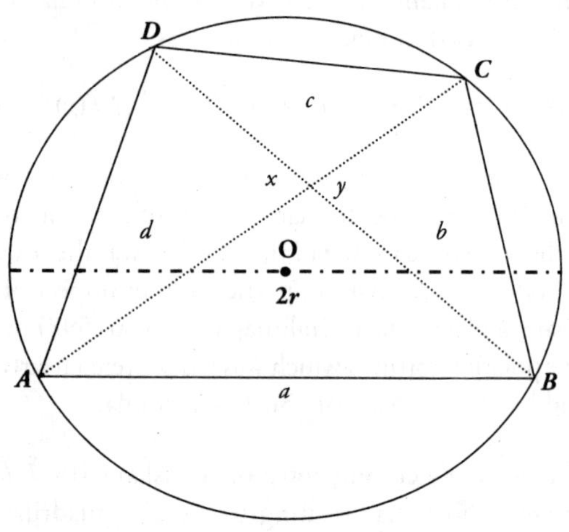

$$r = \sqrt{\frac{(ab+cd)(ac+bd)(ad+bc)}{(a+b+c-d)(b+c+d-a)(c+d+a-b)(d+a+b-c)}}$$

Area of the quadrilateral $ABCD = \sqrt{(s-a)(s-b)(s-c)(s-d)}$ where $s = \frac{1}{2}(a + b + c + d)$

A rationale for these procedures are provided both in the *Yuktibhasa* and the *Kriyakramakari.* These result made their first appearance in Europe in 1782 in the work of l'Huilier (Smith 1925).[9]

6. *The power series for the sine and cosine*, the discussion of which is postponed to a Chapter 7, are usually attributed to Newton. But they are stated in Nilakantha's *Aryabhatiyabhasya* and derived in Jyesthadeva's *Yuktibhasa* and are attributed to Madhava in both texts. In modern notation, and expressed in degrees, the two power series are

$$\sin\theta = \theta - \frac{\theta^3}{3!} + \frac{\theta^5}{5!} - \frac{\theta^7}{7!} + \ldots$$

$$\text{o s}\theta = 1 - \frac{\theta^2}{2!} + \frac{\theta^4}{4!} - \frac{\theta^6}{6!} + \ldots$$

It has been suggested that Madhava may have used the first power series result to construct a sine table of 24 values (i.e., the values of cumulative series obtained by dividing the quadrant of a circle into 24 equal parts) which corresponds to the present-day angular measurement of 3 degrees and 45 minutes. A more plausible explanation is that Madhava developed the 'Suryasiddhanta/Aryabhata' method, discussed in Chapter 5, to arrive at such an accurate table. The values are correct in almost all cases to the 8th or 9th decimal place. These power series make their first appearance in Europe in 1676 in a letter written by Newton to the Secretary of the Royal Society.[10]

7. Nilakantha's *Aryabhatiyabhasya* and Jyesthadeva's *Yuktibhasa* contain the following approximations with the Malayalam '*iti Madhava*' ('thus said Madhava') appearing at the end of the latter text and the Sanskrit '*tatraha Madhavah*' in the former text. The following passage is from the Nilakantha:

> Place the (sine and cosine) chords nearest to the arc whose sine and cosine chords are required and obtain the arc difference to be subtracted or added. For making the correction, 13751 should be divided by twice the arc difference in minutes and the quotient is to be placed as the divisor. Divide the one (say sine) by this (divisor) and add to or subtract from the other (i.e., cosine), according to whether the arc difference is to be added or subtracted. Double (this result) and do as before (i.e., divide by the divisor). Add or subtract the result (so obtained) to or from the first sine or cosine to get the desired sine or cosine chords. (Sarma 1977b)

Expressed in modern symbolic notation, the foregoing rule may be expressed as

$$\sin(\theta + h) \approx \sin\theta + \frac{h}{r}\cos\theta - \frac{h^2}{2r^2}\sin\theta$$

$$\cos(\theta + h) \approx \cos\theta - \frac{h}{r}\sin\theta - \frac{h^2}{2r^2}\cos\theta$$

These results are but special cases of one of the most familiar expansions in mathematics, the Taylor series (named after Brook Taylor, 1685–1731), up to the second power of the small quantity u:

$f(\theta + u) = f(\theta) + uf'(\theta) + (u^2/2)f''(\theta) + \ldots$ where u (in radian measure) $= h/r$

8. In the *Yuktibhasa*, a key result of Kerala mathematics used in the *power series of circular and trigonometric functions* is derived. The demonstration in the *Yuktibhasa* (Sarma 2008: 61–67) follows along the lines explained in the Appendix at the end of this chapter, with the main difference from the *Yuktibhasa* version being that this explanation is given in modern notation. The Appendix also gives the first demonstration in European mathematics by John Wallis and is contained in Proposition I of *Arithmetica Infinitorum* (The Arithmetic of Infinites), published in 1655.

This result was adapted by European mathematicians such as Fermat, Pascal and others in the seventeenth century to evaluate the area under the parabolas $y = x^k$, or, equivalently, calculate $\int x^k \, dx$. Pascal remarked about the utility of this formula:

> Any person at all familiar with the doctrine of indivisibles will perceive the results that one can draw from the above for the determination of curvilinear areas. Nothing is easier, in fact, than to obtain immediately the quadrature of all types of parabolas and the measures of numberless other magnitudes. (Boyer 1959: 113)

Appendix

A Key Result and Contrasting Methods of Proof[11]

In terms of modern notation, the result to be proved may be stated thus:

$$\lim_{n \to 0} \frac{I^k + 2^k \ldots + n^k}{n^{k+1}} = \frac{I}{k+1}$$

The Kerala proof (as explained in Section 4 of Chapter 6 of the *Yuktibhasa*) follows first

The Sum of Natural Numbers (*Mula-Sankalita*)

Let $k = 1$ and $s_1 = 1 + 2 + 3 + \cdots + n$ (1)

Next the following two series are added

$$n + n + \cdots + n + n + n = n^2$$
$$-n - (n-1) - \cdots - 3 - 2 - 1 = -s_1$$

To give $s_1 - n = 0 + 1 + 2 + 3 + \cdots + (n-1) = n^2 - s_1$ (2)

Or, after re-arrangement, $s_1 = {}^1/_2\, n\,(n + 1)$ (3)

As n tends to ∞, $(n + 1)$ can be replaced by n so that $s_1 = {}^1/_2\, n^2$

Hence, $\lim_{n \to \infty} \frac{s_1}{n^2} = \frac{1}{2}$ (4)

Or, equivalently $\lim_{n \to \infty} \frac{1+2+3+. \; +n}{n^2} = \frac{1}{2}$

This proves the result for $k = 1$.

The Sum of the Squares of Natural Numbers (*Varga-Sankalita*)

For $k = 2$, let $s_2 = 1^2 + 2^2 + 3^2 + \ldots + n^2 = 1 \times 1 + 2 \times 2 + 3 \times 3 + \ldots + n \times n$ (5)

Now, as $n \to \infty$, we have by the argument given above

$$1 \times n + 2 \times n + 3 \times n + \ldots + n \times n = s_1 \times n = \tfrac{1}{2} n^3 \quad (6)$$

Subtracting series (5) from (6) gives, as $n \to \infty$,

$$(n-1) \times 1 + (n-2) \times 2 + (n-2) \times 2 + \ldots + (n-n+1) \times (n-1) + (n-n) \times n = \frac{1}{2}n^3 - s_2 \quad (7)$$

The series on the left of the above expression can be written as a triangular sum

$1 + 2 + 3 + \ldots + (n-3) + (n-2) + (n-1)$
$1 + 2 + 3 + \ldots + (n-3) + (n-2)$
$1 + 2 + 3 + \ldots + (n-3)$
$\vdots$
$1 + 2 + 3$
$1 + 2$
1

Now the addition is carried out horizontally (rows). By the previous result as $n \to \infty$, each row can be summed to give

$$\tfrac{1}{2}(n-1)^2 + \tfrac{1}{2}(n-2)^2 + \ldots + 1 = \frac{1}{2}n^3 - s_2$$

As $n \to \infty$, the series on the left side can be taken to be $\frac{1}{2}s_2$.

So we have $\lim_{n\to\infty} \frac{s_2}{n^3} = \frac{1}{3}$

Or, equivalently

$$\lim_{n\to\infty} \frac{1^2 + 2^2 + 3^2 + \ldots + n^2}{n^3} = \tfrac{1}{3}$$

which is the result for $k = 2$.

The sum of third and fourth powers of natural numbers (*Ghana-Sankalita* and *Varga-varga Sankalita* respectively) are derived by a similar argument to above to obtain

$$\frac{s_3}{n^4} = \frac{1}{4} \text{ and } \frac{s_4}{n^5} = \frac{1}{5} \text{ as } n \to \infty$$

Induction is then invoked to establish the general result (*Samaghata-Sankalita*) for all positive integer values of k as

$$\frac{s_k}{n^k} = \frac{1}{k+1} \quad \text{as } n \to \infty$$

Notwithstanding this mode of argument, this is a rigorous but unusual way to show the passage to infinity.

Now contrast the Kerala approach with the earliest work of a similar vein in European mathematics. In Proposition I (of *Arithmetica Infinitorum*, 1656), the English mathematician, Wallis, begins by noting that

$$\frac{0+1}{1+1} = \frac{1}{2}$$

$$\frac{0+1+2}{2+2+2} = \frac{3}{6} = \frac{1}{2}$$

$$\frac{0+1+2+3}{3+3+3+3} = \frac{6}{12} = \frac{1}{2}$$

$$\frac{0+1+2+3+4}{4+4+4+4+4} = \frac{10}{20} = \frac{1}{2}$$

Proceeding in this way, he concludes that this ratio will persist, no matter what the number of terms may be, that is,

$$\frac{0+1+2+3+4+\ldots+n}{n+n+n+n+n+\ldots+n} \text{ will always be equal to } {}^{1}\!/_{2}.$$

Wallis proceeded to deduce that the corresponding ratio of the series $0^n + 1^n + 2^n + 3^n + \ldots$ was $1/(n + 1)$, and that this ratio persists even if n were fractional...

But when he tried to extend his investigations (to negative values of n) the result which emerges proves unintelligible to him. If $n = -1$, as in the hyperbola $y = 1/x$, this ratio becomes $1/0$, which he had already in the *Conic Sections* and elsewhere characterised as infinite. If n were numerically greater than 1, that is, if the curve were of the type $y = 1/x^2$, $1/x^3$, and so on, the consequent of the emergent ratio would be negative. But to him the ratio of a positive quantity to a negative one was meaningless, and his attempts to elucidate it led to the erroneous conclusion that these ratios were *greater than infinite*!

Notes

1. 'Fluxion' was the term used by Isaac Newton for the rate of change (derivative) of a continuously varying quantity, or function, which he called a 'fluent'.
2. For further details regarding this seminal text, see Appendix I of Chapter 6.
3. For further details, see Ramasubramanian et al. (1994).
4. In astronomical calculations, the longitude of a planet is measured along the ecliptic, while its own motion occurs along its own orbit. There is therefore a small difference between these two values. Acyuta Pisarati, for the first time in Indian astronomy, gave a formula for the reduction to the ecliptic in the case of the moon in his work, *Sputanirnaya* and a simpler version of that formula in another work, *Uparagakriyakrama* (Procedures for Computing Eclipses). In Europe, a similar formula for the reduction to the ecliptic was formulated by the Danish astronomer, Tycho Brahe in his *Astronomiae instaurtae Proggymnasmata* which was published a year after his death in 1602.
5. Or more precisely, as we sum more terms, the difference between 1/3 and sums of powers of 1/4 as given by the term in the square bracket becomes extremely small, but never zero. Only when we take all the terms of the infinite series together do we obtain the equality

$$1/4 + 1/4^2 + 1/4^3 + \ldots = 1/3$$

6. The term 'triangular' originates with the Greeks for whom there were certain numbers which could be represented with a triangular array of dots:

1 = • **3** = • • • **6** = • • • • • • **10** = • • • • • • • • • • etc.

Triangular numbers can also be seen as the sums of consecutive natural numbers beginning with 1 = 1, 3 = 1 + 2, 6 = 1 + 2 + 3, 10 = 1 + 2 + 3 + 4, 15 = 1 + 2 + 3 + 4 + 5, . . . , + T_n

where $T_n = 1 + 2 + 3 + \cdots + (n-1) + n$
$T_n = n + (n-1) + (n-2) + \cdots + 2 + 1$
Adding gives $2T_n = (n+1) + (n+1) + (n+1) + \cdots + (n+1) + (n+1)$

There are n groups of $(n + 1)$, so we see that $2T_n = n(n+1)$
or $T_n = n(n+1)/2$

7. Expressed mathematically, the formula for the sum of sums, given the notation of the previous note, is

$$T_1 + T_2 + T_3 + \cdots + T_n = n\frac{(n+1)}{2}.\frac{(n+2)}{3} \text{ where } T_r = \frac{r(r+1)}{2}$$

8. A cyclic quadrilateral is a quadrilateral inscribed in a circle. The circum-radius of a quadrilateral is the radius of the circle in which the quadrilateral is inscribed.
9. In Chapter 7 of the *Yuktibhasa* (Sarma 2008: 109–33), there is a more extensive discussion of the properties of cyclical quadrilaterals based on earlier work and notably that of Paramesvara.
10. An intriguing question is why the sine series was needed in Europe when all work there relating to trigonometric tables was expressed in degrees rather than radians.
11. The discussion of the Kerala approach in this Appendix is based on Sarma's translation of the *Yuktibhasa* (2008: 61–64, 192–95).

5

Indian Trigonometry: From Ancient Beginnings to Nilakantha*

A Preliminary Note

The origins of trigonometry as a discipline may be traced back to Hipparchus of Nicaea (fl. 150 BC). An aid to the study of astronomy/astrology, the basic problem to be tackled was to estimate, for a given arc of a circle, the length of the chord that connects the endpoints of that arc. It was soon noticed that the chord length depends on both the length of the arc and the radius of the circle. For the ancients (Greeks as well as the others), an angle of 90 degrees was not a right angle (as we understand it today) but the quarter of the circumference of a circle. Or generally,

* This chapter is based on some primary data research undertaken by Dr Mallayya when he was a research associate with a project on Kerala mathematics funded by Art and Humanities Reseach Board (AHRB), United Kingdom. The output took the form of a survey (unpublished) entitled 'Trignometric Sines and Sine Tables in India'. This will be made available eventually in the archives of the research project. Some of the ideas contained in this chapter also appeared in a paper entitled 'Kerala Mathematics: Motivation, Rationale and Method' presented at a workshop and then published in an edited volume of the proceedings authored by Mallayya and Joseph (2009b). The author owes a considerable debt to Dr Mallayya for his linguistic and mathematical expertise as well as his deep knowledge of Kerala mathematical history that he brought to the project.

'degrees' was more a measure of the length of an arc rather than the size of an angle. Thus, for a given circle of circumference of 360 degrees, the radius of the circle was $360/2\pi = 57.2957$, if π was approximated by the Aryabhatan value of 3.1416 as given in Verse 10 of the *Aryabhatiya* and discussed in Chapter 3. Or if the circumference was measured in minutes (60 minutes = 1 degree), then the total circumference is $360 \times 60 = 21600$ minutes and the corresponding radius is 3437.7467... or approximately 3438 minutes. This was known fairly early in Indian trigonometry with the radius of 3438 minutes taken as a standard measure in the construction of trigonometric tables.[1]

One of the earliest trigonometric tables constructed contained values for the length of the chord for a given arc (normally denoted by crd α). In terms of modern notation and as shown in Figure 5.1, the chord for a given arc is twice the sine of half the angle multiplied by the radius of the circle taken here to be 3438. On the basis of this relationship, a table for chords can be constructed.[2]

Figure 5.1

The Length of a Chord

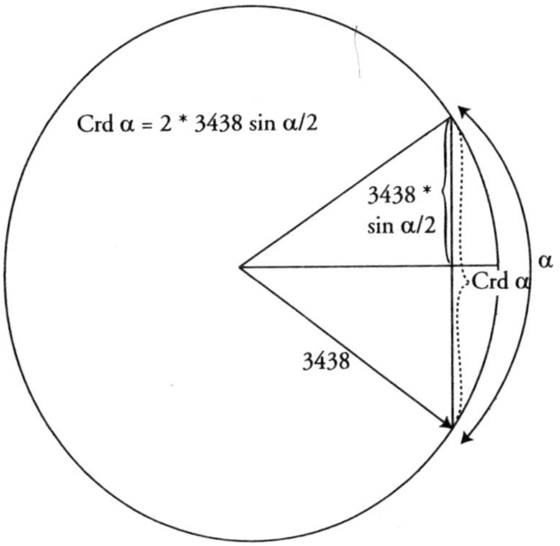

Source: Bressoud (2002: 2).

Traditional Indian Trigonometry: Introduction and Terminology

Figure 5.2 represents a circle of radius *R*. The following terminology and relationships may be deduced by reference to this diagram.

1. The arc *PQ* of the circle was known as *capa*.
2. The chord *PQ* was known as *samastajya*.
3. The half-chord *PM* (or Indian Sine) was known as *ardhajya* or *jyardha* or *bhujajya* and often shortened to *jya*.
4. The relationship between the Indian Sine and the modern sine is a simple one:
   Indian Sine (*jya*) $= s = R\theta = R\text{Sin}\theta$ (where $s = R\theta =$ arc of angular measure and Sine $\theta =$ modern sine).
   [Note: The Indian Sine, often denoted as Sine with a capital 'S', is *not* a ratio but a linear measure.]
5. *OM* was called *kotijya* (or *kojya*) and equals $R\text{Cos}\theta$ where $\cos\theta$ is the modern cosine.

Figure 5.2

Basic Concepts in Indian Trigonometry

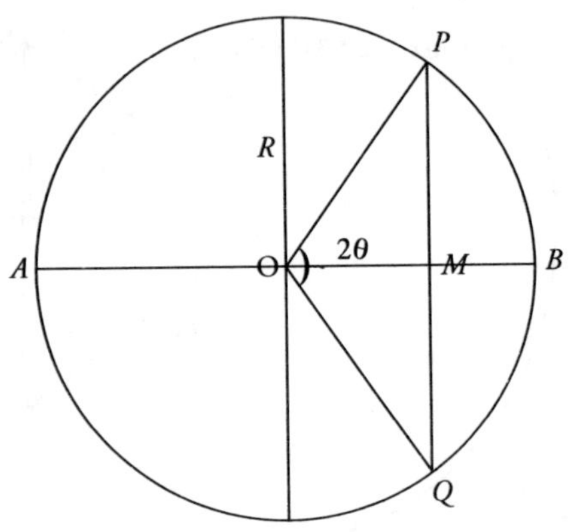

6. Now $MB = OB - OM = R - kojya = R - R\text{Cos}\,\theta = R(1 - \cos\theta)$. This was known as *sara* or *utkramajya* which is the *R*versine θ.
7. The radius R of the circle was referred variously as *vyasadala* or *vyasardha or* even *sinjini*. However, in later texts, such as the *Tantrasangraha* of Nilakantha, *sinjini* is the Indian Sine or RSine.

The Work of Aryabhata I (b. AD 476)

The construction of a set of successive RSines (or *jyas*) and their differences for purposes of interpolation (or for calculating the intermediate values) has been a common feature of Indian astronomical texts from the time of Aryabhata I. These values were then used for accurate computations of planetary positions. As mentioned earlier, taking the circumference of a circle in angular measure of $360° \times 60 = 21600$ minutes led to an estimated radial value of approximately 3438′, using the Aryabhatan value of 3.1416 as the ratio of circumference to diameter. The circumference was then divided into 24 equal parts so that each part covered an angular measure of 3°45′ or 225′.

In the *Aryabhatiya*, as we have discussed in Chapter 3, Aryabhata gave a method for computing of RSines geometrically which yielded a table of 24 RSine-differences at intervals of 225′. The geometrical method is explained in Verse 11 of the chapter on *Ganitapada* and stated as: Divide a quadrant of the circumference of a circle such that the resulting (right-angled) triangles and quadrilaterals will produce as many RSines (*jyardhas*) as desired for any given half diameter.

In the next verse, the method for constructing a set of RSine-differences is stated tersely: Divide the RSine of the first arc by itself and reduce it by the quotient to obtain the second RSine-difference. That first RSine diminished by all the quotients obtained by dividing each of the preceding RSines by the first RSine gives the remaining RSine-differences.[3]

Expressed in modern notation, the first RSine-difference (d_1) is given by

$$d_1 = \text{RSin}\theta = R\theta = h = 225' \text{ since } \sin\theta \approx \theta \text{ if } \theta \text{ is small (note } \sin 0° = 0)$$

The second RSine-difference d_2 is given by:

$$d_2 = R\text{Sin}\theta - \frac{R\text{Sin}\theta}{R\text{Sin}\theta} = R\text{Sin}\theta - 1 = 225 - 1 = 224'$$

The third RSine-difference d_3 is given by the formula

$$d_3 = R\text{Sin}\theta - \left[\frac{R\text{Sin}\theta}{R\text{Sin}\theta} + \frac{R\text{Sin}2\theta}{R\text{Sin}\theta}\right]$$

The fourth RSine-difference d_4 is computed using the formula

$$d_4 = R\text{Sin}\theta - \left[\frac{R\text{Sin}\theta}{R\text{Sin}\theta} + \frac{R\text{Sin}2\theta}{R\text{Sin}\theta} + \frac{R\text{Sin}3\theta}{R\text{Sin}\theta}\right] \ldots \text{ and so on}$$

So that in general, the $(n+1)^{\text{th}}$ RSine-difference d_{n+1} is given by

$$d_{n+1} = R\text{Sin}\theta - \left[\frac{R\text{Sin}\theta}{R\text{Sin}\theta} + \frac{R\text{Sin}2\theta}{R\text{Sin}\theta} + \frac{R\text{Sin}3\theta}{R\text{Sin}\theta} + \ldots + \frac{R\text{Sin}(n\theta)}{R\text{Sin}\theta}\right]$$

That is,

$$R\text{Sin}\,[(n+1)\theta] - R\text{Sin}\,(n\theta) = R\text{Sin}\theta - \left[\frac{R\text{Sin}\theta}{R\text{Sin}\theta} + \frac{R\text{Sin}2\theta}{R\text{Sin}\theta} + \frac{R\text{Sin}3\theta}{R\text{Sin}\theta} + \ldots + \frac{R\text{Sin}(n\theta)}{R\text{Sin}\theta}\right]$$

Or,

$$R\text{Sin}\,(n+1)\theta = R\text{Sin}\theta - \left[\frac{R\text{Sin}\theta + R\text{Sin}2\theta + R\text{Sin}3\theta + \ldots + R\text{Sin}\,n\theta}{R\text{Sin}\theta}\right] + R\text{Sin}\,n\theta$$

This expression was then used to generate the RSine-differences successively from the preceding ones and also to find the RSines successively.

The *Sūryasiddhanta* (c. AD 400; Shukla 1957: ii,15–22), an earlier text than *Aryabhatiya*, gives the same formula along with a list of 24 RSines, the first being stated as an eighth part of a *rasi*.[4] Since a *rasi* comprises 30° an eighth part of a *rasi* is equal to 3° 45′ or 225 minutes. The 24 RSines are then obtained successively in the following manner.

> The eighth part of a *rasi* is the first *R*Sine. Divide this by itself, subtract the quotient and add to itself. This is the second. By the first, divide each of the *R*Sines obtained in order, subtract the sum of all the quotients from the first and add the last of the *R*Sine obtained. The twenty four *R*Sines are thus obtained successively. (Shukla 1957)

Elaborating on this *Sūryasiddhanta* rule, the Kerala mathematician Paramesvara (fl. 1430) gave a formula for computing the RSine-differences successively from the preceding RSine-differences according to which the first RSine divided by itself and diminished by the quotient gives the second RSine-difference. The other RSine-differences are obtained successively by subtracting from the preceding difference, the quotient of division of sum of all the preceding RSine-differences by the first RSine.

In other words, if d_1, d_2, d_3, ... are the successive RSine-differences then

$$d_1 = R\text{Sin}\theta\left(\frac{30^\circ}{8} = 225'\right)$$

$$d_2 = d_1 - \frac{d_1}{R\text{Sin}\theta}$$

$$d_3 = d_2 - \frac{d_1 + d_2}{R\text{Sin}\theta}$$

Or in general,

$$d_{n+1} = d_n - \frac{d_1 + d_2 + d_3 + \ldots + d_n}{R\text{Sin}\theta}$$

Since $d_1 = R\text{Sin}\ \theta$ and $d_k = R\text{Sin}\ k\theta - R\text{Sin}\ (k-1)\theta$ for $k = 1, 2, 3, \ldots, n$, the above implies

$$d_{n+1} = d_n - \frac{R\text{Sin}\theta + \{R\text{Sin}2\theta - R\text{Sin}\theta\} + \{R\text{Sin}3\theta - R\text{Sin}2\theta\} + \ \ldots\ + \{R\text{Sin}\ n\theta - R\text{Sin}(n-1)\theta\}}{R\text{Sin}\theta}$$

Or simplifying the expression, we get

$$d_{n+1} = d_n - \frac{R\text{Sin}(n\theta)}{R\text{Sin}\theta}$$

This simple formula could then be used to compute successively the RSine-differences and also the RSines.

The initial value of *jya* ($R\text{Sin}\theta$) given by Aryabhaṭa is 225 minutes. He lists his successive values of RSine-differences (d_1, d_2, d_3, ..., d_{24}) values as follows:

225,	224,	222,	219,	215,	210,
205,	199,	191,	183,	174,	164,
154,	143,	131,	119,	106,	93,
79,	65,	51,	37,	22,	7

For these differences, the *Sūryasiddhanta*, using identical methods, lists the values of 24 RSines at intervals of 225 minutes as follows:

225, 449, 671;
890, 1105, 1315;
1520, 1719;
1910, 2093;
2267, 2431;
2585, 2728;
2859, 2978;
3084, 3177;
3256, 3321;
3372, 3409;
3431, 3438;

Table 5.1 presents a table of RSines and RSine-differences given by Aryabhata computed for arcs at interval of 225′. A comparison is made between these values (after adjustments obtained by dividing the RSine values by the value of $R = 3438$) and the corresponding modern values. The remarkable accuracy of a table computed from methods that are at least 1,500 years old is very striking.

The Work of Varahamihira (fl. AD 500)

A near contemporary of Aryabhata used a different but interesting method of computing sine values. Varahamihira takes $\sqrt{10}$ as the value of circumference–diameter ratio[5] and assumes the radius to be 2° (= 120′) for computing the RSines of arcs at interval of 3°45′ (= 225′) of angular measure. For a circumference of angular measure 21,600 minutes, the radius will be 3415.259873 instead of 3438, the one used most often in Indian astronomy. However, this variation in the value of R does not affect the RSine computation by Varahamihira in any way because the RSines he derives does not depend on the ratio of circumference to diameter.

Varahamihira provides a detailed discussion of RSines and construction of trigonometric tables in his famous astronomical treatise *Panchasidhantika*. In it appears: 'The square of half diameter is termed *dhruva*. Its fourth part *mesa* (is of first *rasi*); *dhruvakarani* diminished

Table 5.1

Aryabhata's Sine Table

No: k	Arc in angular measure, $k\theta'$	RSine-diff, d_k	RSine values J_k	Sine values $S_k = J_k/R$ ($R = 3438'$)	Modern value of sines s_k	Computation error, $\epsilon_k = S_k - s_k$
1	225	225	225	0.065445026	0.065403129	0.00004190
2	450	224	449	0.130599186	0.130526192	0.00007299
3	675	222	671	0.195171611	0.195090322	0.00008129
4	900	219	890	0.258871437	0.258819045	0.00005239
5	1125	215	1105	0.321407795	0.321439465	−0.00003167
6	1350	210	1315	0.38248982	0.382683432	−0.00019361
7	1575	205	1520	0.44211751	0.442288769	−0.00017126
8	1800	199	1719	0.5	0.5	0
9	2025	191	1910	0.555555556	0.555570233	−0.00001468
10	2250	183	2093	0.608784177	0.608761429	0.00002275
11	2475	174	2267	0.659394997	0.659345815	0.00004918
12	2700	164	2431	0.70709715	0.707106781	−0.00000963
13	2925	154	2585	0.751890634	0.751839807	0.00005083
14	3150	143	2728	0.793484584	0.79335334	0.00013124
15	3375	131	2859	0.831588133	0.831469612	0.00011852
16	3600	119	2978	0.86620128	0.866025403	0.00017588
17	3825	106	3084	0.897033159	0.896872741	0.00016042
18	4050	93	3177	0.92408377	0.923879532	0.00020424
19	4275	79	3256	0.947062245	0.946930129	0.00013212
20	4500	65	3321	0.965968586	0.965925826	0.00004276
21	4725	51	3372	0.980802792	0.98078528	0.00001751
22	4950	37	3409	0.991564863	0.991444861	0.00012000
23	5175	22	3431	0.997963933	0.997858923	0.00010501
24	5400	7	3438	1	1	0

Source: Mallayya (2009: 13).

by that of *mesa* is of two *rasi*. The square root (of a *mesakarani*) is the RSine.[6]

Since the diameter is 240 minutes, *dhruva* (or square of radius) = $R^2 = 14400$, and one *rasi* is 30°, so RSine of one *rasi* is $R\text{Sin}30°$.

Now *mesakarani* $= \dfrac{R^2}{4} = 3600$ or $(R\text{Sin}30°)^2 = \dfrac{R^2}{4} = 3600$

According to the above-stated rule

$$\text{The square of RSine of two } rasi = dhruvakarani - mesakarani$$
$$= R^2 - (R\text{Sin } 30°)^2$$
$$= 14400 - 3600 = 10800$$
$$\text{Or } (R\text{Sin } 60°)^2 = R^2 - (R\text{Sin } 30°)^2 = 10800.$$

Taking square roots,

$$R = \sqrt{14400} = 120'$$

$$R\text{Sin}30° = \sqrt{\frac{R^2}{4}} = \frac{R}{2} = 60'$$

$$R\text{Sin } 60° = \sqrt{R^2 - \left(R\text{Sin } 30°\right)^2}$$

$$= \sqrt{R^2 - \frac{R^2}{4}} \quad \text{(which is equal to } R\frac{\sqrt{3}}{2}\text{)}$$

$$= \sqrt{10800}$$
$$= 103.9230485'$$
$$\approx 103'55''.$$

For other tabular sines, the following methods were used.

$$\sin^2\theta = \frac{1}{4}[\sin^2 2\theta + \{120 - \sin(90 - 2\theta)\}^2]$$

and

$\sin^2\theta = 60[120 - \sin(90 - 2\theta)]$ where $R = 120'$ and the sines are in minutes.

Using the results given above, a table of the 24 RSines was computed and is given in Figure 5.2. Corresponding values of modern sines are then computed from them by dividing by R (taken as 120). A comparison may then be made with modern values showing the error in each value computed. Note that the accuracy is even more remarkable than Aryabhata's sine table.

Table 5.2

Varahamihira's Table of Sine-Differences

No. k	Arc = $k\theta'$ (in angular measure)	RSine-differences	RSines as given in Table 5.1 (J_k)	RSine-differences (computed from RSines) ($J_{k+1}-J_k$)
1	225	7'51"	7'51"	7'51"
2	450	7'49"	15'40"	7'49"
3	675	7'45"	23'25"	7'45"
4	900	7'39"	31'4"	7'39"
5	1125	7'30"	38'34"	7'30"
6	1350	7'22"	45'56"	7'22"
7	1575	7'9"	53'5"	7'9"
8	1800	6'55"	60'0"	6'55"
9	2025	6'40"	66'40"	6'40"
10	2250	6'23"	73'3"	6'23"
11	2475	6'4"	79'7"	6'4"
12	2700	5'44"	84'51"	5'44"
13	2925	5'22"	90'13"	5'22"
14	3150	4'59"	95'12"	4'59"
15	3375	4'34"	99'46"	4'34"
16	3600	4'9"	103'55"	4'9"
17	3825	3'42"	107'37"	3'42"
18	4050	3'15"	110'52"	3'15"
19	4275	2'45"	113'37"	2'45"
20	4500	2'18"	115'55"	2'18"
21	4725	1'47"	117'42"	1'47"
22	4950	1'17"	118'59"	1'17"
23	5175	0'45"	119'44"	0'45"
24	5400	0'16"	120'0"	0'16"

Source: Mallayya (2009: 22).

The Work of Bhaskara I (fl. AD 600)

Bhaskara I, in his *Mahabhaskariya*, gave two methods for computation of RSines. One is without using RSine-differences and the other using RSine-differences. It is the first of the methods that was truly innovative. It has already been examined in the Appendix to Chapter 3. A sine table was constructed applying this rule.

The other rule given by Bhaskara I gives RSines and Rversed sines of arcs using RSine-difference table and it is based on a simple interpolation technique using the concept of proportion. The rule may be translated as follows:

> The given arc expressed in minutes is divided by 225. The quotient obtained is the number of *R*Sine-differences to be taken completely (for computing sum). Again, the remainder multiplied by the current (difference) is divided by 225. This added to the previous sum (of *R*Sine-differences) gives the *R*Sines of the given arc (<90°), or (likewise) *R*versed Sines.[7]

According to this rule, if $\frac{\text{arc}}{225} = n + \frac{\theta}{225}$ where the arc is in minutes, the quotient of division of the arc by 225 is n and the remainder is θ, then the *jya* of the given arc is equal to:

$$\text{Sum of the first } n \text{ RSine-differences} + \frac{\text{remainder} \times \text{current difference}}{225} = \sum_{i=1}^{n} \mathrm{d}_i + \frac{\theta d_{n+1}}{225}$$

(where $d_{k+1} = R\text{Sin}\,(k+1)h - R\text{Sin}\,(kh)$ and h is the interval of differencing 225 in the table of RSines).[8]

Using this formula, the values of RSines can be interpolated for different $u = \frac{\theta}{h} < 1$ (since $\theta < h$), and a table of RSines (and also *R*Versines) may be prepared for values of RSine-differences up to 90°.[9]

To find RSines of arcs having angular measure greater than 90°, Bhaskara I prescribes the following formulae in his *Mahabhaskariya* (iv-2).

$$\begin{aligned} R\text{Sin}(90^\circ + \theta) &= R\text{Sin}90^\circ - R\text{Versin}\theta \\ R\text{Sin}(180^\circ + \theta) &= R\text{Sin}90^\circ - R\text{Versin}90^\circ - R\text{Sin}\theta = -R\text{Sin}\theta \\ R\text{Sin}(270^\circ + \theta) &= R\text{Sin}90^\circ - R\text{Versin}90^\circ - R\text{Sin}90^\circ + R\text{Versin}\theta \\ &= -R\text{Sin}90^\circ + R\text{Versin}\theta \text{ for } \theta < 90^\circ. \end{aligned}$$

While Bhaskara I provides this interpolation formula up to first order term, Brahmagupta extends it to second order term in his works *Dhyanagrahopadesadhyaya* (before AD 628) and *Khandakhadyaka* (AD 665).

The Work of Brahmagupta (b. AD 598)

Brahmagupta's rule is as follows:

> Multiply half the difference of the *gata khanda* (i.e., tabular difference d_g passed over) and *bhogya khanda* (i.e., the difference d_b to be passed over) by the residual arc (θ in minutes) and divide by 900′ minutes. The result is added to and subtracted from half the sum (of d_g and d_b) according to whether this half sum is less than or greater than the tabular difference to be crossed. The result obtained is the true functional difference to be crossed.

According to this a better estimate of the functional difference or true functional difference to be crossed is

$$d=\frac{1}{2}\left(d_g+d_b\right)+\frac{1}{2}\left(d_g-d_b\right)\frac{\theta}{h}, \text{ or } \quad d=\frac{1}{2}\left(d_g+d_b\right)-\frac{1}{2}\left(d_g-d_b\right)\frac{\theta}{h},$$

depending on whether $\frac{1}{2}\left(d_g d_b\right) <$ or $> d_b$, or $d_g <$ or $> d_b$ where $h =$ 900′ and $d_b = d_{g+1}$.

The interpolated value is $f(x + hu) = f(x) + u\,d$ where d is as given above and $\theta = hu$.

Hence, $f(x+hu)=f(x)+\frac{u}{2}(d_g+d_b)-\frac{u^2}{2}(d_g-d_b)$ if $d_g > d_b$

This gives $f(x+hu)=f(x)+\frac{u}{2}(d_g+d_{g+1})-\frac{u^2}{2}(d_g-d_{g+1})$

From this we get

$$f(x+hu)=f(x)+\frac{u}{2}\{\Delta f(x-h)+\Delta f(x)\}-\frac{u^2}{2}\{\Delta f(x-h)-\Delta f(x)\}$$

Or $f(x+hu)=f(x)+\frac{u}{2}\{\Delta f(x-h)+\Delta f(x)\}+\frac{u^2}{2}\{\Delta^2 f(x-h)\}$

where $\Delta f(x)=f(x+h)-f(x)$.

This is the Newton–Sterling's interpolation formula up to second order term.

Table 5.3 is constructed using the Brahmagupta's method at intervals of 15° (or 900 minutes) with the RSine-differences of order up to two. The value of R used by Brahmagupta was 150.

Table 5.3

Brahmagupta's Table of Sines

k	Arc = $k\,\theta$	RSines	RSine-differences	
			First order	Second order
1	15° = 900′	39	39	
2	30° = 1,800′	75	36	−3
3	45° = 2,700′	106	31	−5
4	60° = 3,600′	130	24	−7
5	75° = 4,500′	145	15	−9
6	90° = 5,400′	150	5	−10

Source: Mallayya (2009: 30).

The Work of Vaṭesvara (b. AD 880)

In the first chapter of the *Vaṭesvara Siddhanta* the author writes:

> After the lapse of 802 years from the beginning of the *saka* era, I was born; and 24 (years) after my birth, this *Siddhanta* was composed by me by with the blessings of all planets.

From this information we conclude that Vatesvara was born in 802 Saka era (or AD 880) and composed his work 24 years later in AD 904.

Well versed in the work of the Aryabhatan School, Vatesvara shows considerable skill in trigonometric computations. In verses 2 to 51 of his *Siddhanta*, he gives a list of the values of 96 RSines and 96 RVersed Sines at intervals of 56.25 minutes. This list is interposed with verses giving various relationships between RSines, RCosines and RVersed Sines in various quadrants, several methods for computing the desired RSines from given arc and tabular values of RSines. Several forms of first and second order interpolation techniques and several inverse interpolation methods are given in Verses 55–92 for finding the desired arc from the given RSine and tabular values.

On the basis of this information, tables of 96 values of RSines and RCosines can be constructed and a comparision made of these values with corresponding modern values.[10] The value of R used by Vatesvara for computation of the tables was 3437′ 44″ and not 3438′. Earlier, the Kerala astronomer Govindasvamin (AD 800–850) used the even more accurate value 3437′ 44″ 19‴ for *R* for sine computation. Note that the value of *R* used by Madhava (c. 1340–1425), the founder of the Kerala School of mathematics and astronomy, was 3437′ 44″ 48‴. Whether the Govindasvamin influenced Vatesvara or Madhava is a matter of conjecture and certainly worth investigation.

Vatesvara's treatment of trigonometric relationships remains one of the most comprehensive and innovative achievements of early Indian trigonometry.

The Work of Bhaskara II (b. AD 1114)

Bhaskara II in the *Siddhantasiromaṇi* (1150) gives similar rule as his predecessors for interpolating intermediate RSine values and Munisvara in his commentary *Marici* (1635) on Bhaskara's work modifies the rule to attain greater accuracy. According to Bhaskara II:

> Multiply the difference of the crossed and to be crossed differences by the residual arc, divide by twenty and subtract it from or add to half the sum of the differences crossed and to be crossed. This will be the true difference to be crossed for getting *kramajya* or *utkramajya* respectively.

Or, the true difference to be crossed over is given by:

$$d = \frac{1}{2}\left(d_g + d_b\right) - \left(d_g - d_b\right)\frac{\theta}{20} \text{ for computing RSines}$$

and

$$d = \frac{1}{2}\left(d_g + d_b\right) + \left(d_g - d_b\right)\frac{\theta}{20} \text{ for computing RVersine}$$

where d_g is the tabular difference passed over and d_b is the tabular difference to be crossed.[11]

Munisvara employs an iterative procedure numerically for attaining the functional difference with desired degree of accuracy using which RSines can be interpolated with the desired degree of accuracy.

The interpolated functional value is given by $f(x + hu) = f(x) + \mu\,\delta$ where δ is the current functional difference with desired degree of accuracy and $u = \frac{\theta}{h}$.

Suppose Munisvara's successive approximations to δ are denoted by, $\delta^{(1)}, \delta^{(2)}, \delta^{(3)}, \ldots$ If the number of iterations n becomes larger and larger, the desired degree of accuracy is attained when $\delta^{(n+1)} = \delta^{(n)}$. This value is taken as the value δ of the current functional difference with the desired degree of accuracy.

The true tabular difference d for interpolating RSines according to Bhaskara II is given by

$$d = \frac{1}{2}\left(d_g + d_b\right) - \frac{1}{2}\left(d_g - d_b\right)\frac{\theta}{h}, \quad \text{where } h = 600'$$

$$= \frac{1}{2}\left(d_g + d_b\right) - \frac{\theta}{2h}d_g + \frac{\theta}{2h}d_b$$ where d_g is the tabular difference crossed,

d_b is the tabular difference to be crossed and θ is the residual arc.

The iteration process is initiated by taking as the first approximation to δ, $\delta^{(1)} = d$. Now using this, the other successive approximations to δ, for example, the second approximation $\delta^{(2)}$, third approximation $\delta^{(3)}$, fourth approximation $\delta^{(4)}$, and so on, may be computed using the formula

$$\delta^{(n+1)} = \frac{1}{2}\left(d_g + d_b\right) - \frac{\theta}{2h}d_g + \frac{\theta}{2h}\delta^{(n)}$$

As n becomes larger, the desired accuracy will be attained at some stage when $\delta^{(n+1)} \approx \delta^{(n)}$ and this value is taken as the value of δ.
Hence when $n \to \infty$,

$$\delta = \frac{1}{2}\left(d_g + d_b\right)\frac{\theta}{2h}d_g + \frac{\theta}{2h}\delta$$

Or,

$$\delta\left(1 - \frac{\theta}{2h}\right) = \frac{1}{2}\left(d_g + d_b\right)\frac{\theta}{2h}d_g$$

From this we get

$$\delta = \frac{(h-\theta)d_g + hd_b}{2h-\theta}$$

Substituting this value of δ in $f(x + hu) = f(x) + \mu\delta$, we get

$$f(x+hu) = f(x) + u\left\{\frac{(h-\theta)d_g + h.d_b}{2h-\theta}\right\}$$

Using this interpolated functional value one can obtain a desired degree of accuracy. In other words, RSine of arc $x + hu$ can be computed within any desired degree of accuracy.

The *Golasara* of Nilakantha

Over the long period under discussion, Indian mathematicians continued to make significant contributions to trigonometry under the caption '*jyotpatti*' (*jya* + *utpatti* = source of RSines). This branch of mathematics evolved from astronomical needs, such as computation of latitudes or of position of planets, their movements, and so on. As Nilakantha pointed out in his *Golasara*,[12] while explaining the concept of *jya* (or RSines), that he was computing sines and cosines because they were required for a discussion of the motion of planets in their respective orbits on the stellar sphere.[13] As a case study of the advances made in trigonometry by the Kerala School, consider briefly Nilakantha's method of computing sine tables as explained in his *Golasara*.[14]

The *Golasara* shows two methods of computing sine tables. Only the first method intended to find the value of first RSine, using sine tables of length 3×2^m (for $m = 0, 1, 2, 3, 4, 5, \ldots$) will be shown here. The relevant Verses 6–14 have been translated by Mallayya (2009) as follows:

> In a circle, make half its diameter (*vyasardham*) the base (*bhumi*) and the other two sides (*bhujas*) equal to it. From the join of the two sides (*bhujayoga*) a perpendicular [is drawn such that], the two base segments (*abadhas*) are equivalent to half the side (*bahvardham*).
>
> Or, the base segment here is the RSine (*ardhajya*) of the residual arc (*sistacapa*) of the quadrantal arc (*paridheh pada* means one fourth of the circumference). Hence the RSine of a *rasi* (a twelfth part of the circumference, which is equal to an arc of angular measure 30°) is equal to half of the half diameter (*vyasardhadalam*).
>
> Its RCosine (*koti*) is the perpendicular (*lamba*). The hypotenuse (*karna*) of (the triangle with sides) the two half chords is the half of diameter.
>
> The half diameter minus the RCosine is the RSine's arrow or the RVersine (*bahorbana*). Then from their hypotenuse, RSines and the like are to be found skillfully again and again in the circle.

> The square root of one and one third part (*satryamsa*) of the RVersine augmented by the square of the RSine is the approximate arc.
>
> In this manner, some definite portions of the circumference are to be formed. Of these arcs, the RSine differences are mutually dependent on the two RSines in their midst, also the next difference in the same direction (ie, in the same sense) on the RSine at the junction of the two arcs. The circumference of circle is also assumed to be entirely subdivided into minutes of arcs.
>
> If the diameter is equal to 113 (*visvaika samo vyasah*), then the circumference will be tending to 355 (*aagunarthaban*).
>
> That whatsoever is got here of whatever form in whatever manner by whatever means, then, that any other(s) obtained, in that manner by that means will be similar and of the same form.
>
> Multiply this half chord (*jyardha*, RSine) by the double of the very last difference and divide by the radius. Add the result to the difference and subtract from the RSine to get the preceeding RSine.
>
> The RSine of any desired deficit or excess arc is found by using the hypotenuse, base and perpendicular, and so on. (Mallayya 2004: 40–55)

The method here is to find the RSines geometrically using a circle of radius R. In Figure 5.3, R is to be taken as the base BC of an equilateral triangle ABC whose other two sides AC and AB are equal to the radius R. Drop the perpendicular AL from the join A of the two sides AB and AC where C is the centre of the circle.

Then, the *abadhas* (or base segments) are CL and BL, each equal to half the radius (R)

$$\text{That is, } CL = BL = \frac{R}{2}$$

But $CL = AM = R\text{Sin } 30° =$ half the chord AB'

Or $R\text{Sin}30° = \frac{R}{2}$ or RSine of one *rasi* $= \frac{R}{2}$ (Note that the modern $\sin 30° = \frac{1}{2}$).

Now $AL =$ half the chord $AA' = R\text{Sin}60°$ and $CM = AL = R\text{Cos}30°$, the *koti* of 30° arc

Thus $R\text{Sin}60°$ or $R\text{Cos}30° = AL$ or $CM = \sqrt{R^2 - \left(\frac{R}{2}\right)^2} = R\sqrt{\frac{3}{4}}$

Figure 5.3

A Geometric Evaluation of RSines

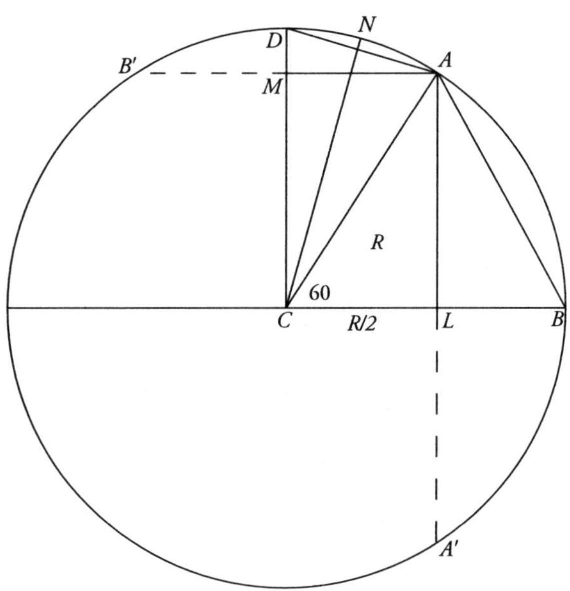

And

$$R\text{Versine of } 30° \text{ arc is given by } MD = CD - CM = R - R\text{Cos}30° = R(1 - \text{Cos}30°)$$

Thus Radius – RCosine gives RVersine

Now from the hypotenuse of these two (that is, the RSine of 30° and the RVersine of 30°), the RSine of 15°, $7\frac{1}{2}°$, … and so on can be found by repeated application of the same process on the circle.

$$AM = R\text{Sin}30° = \frac{R}{2};\ MD = \text{RVersine } 30° = R\,(1 - \text{Cos}30°)$$

$$AD^2 = AM^2 + MD^2$$

$$R\text{Sin } 15° = DN = \text{half the chord } AD = \frac{1}{2}AD = \frac{1}{2}\sqrt{AM^2 + MD^2}$$

$$= \frac{1}{2}\sqrt{\left(\frac{R}{2}\right)^2 + \{R(1 - \text{Cos}30°)\}^2}$$

Or $$=\frac{1}{2}\sqrt{\left(\frac{R}{2}\right)^2+\{R(1-\text{Sin}60^\circ)\}^2}$$ in other terms.

Thus

$$R\text{Sin}15^\circ=\frac{1}{2}\sqrt{\left(\frac{R}{2}\right)^2+R^2\left(1-\frac{\sqrt{3}}{2}\right)^2}$$

$$=\frac{1}{2}\sqrt{\frac{R^2}{4}+R^2\left(1+\frac{3}{4}-\sqrt{3}\right)}$$

$$=\frac{R}{2}\sqrt{2-\sqrt{3}}$$

Or $$\text{Sin } 15^\circ=\frac{1}{2}\sqrt{2-\sqrt{3}}$$

Now $$R\text{Cos}15^\circ = CN = \sqrt{CD^2-DN^2} = \sqrt{R^2-\frac{R^2}{4}\left(2-\sqrt{3}\right)} = \frac{R}{2}\sqrt{2+\sqrt{3}}$$

Similarly, RSine and RCosine of $7\frac{1}{2}^\circ$, $3\frac{3}{4}^\circ$ ... and so on, can be determined geometrically. The general formula embedded in this formulation is

$$R\text{Sin}\left(\frac{\theta}{2}\right)=\frac{1}{2}\sqrt{(R\text{Sin}\theta)^2+(R\text{Versin}\theta)^2}=\frac{1}{2}\sqrt{(R\text{Sin}\theta)^2+\{R(1-\text{Cos}\theta)\}^2}$$

which is true since the right side of this expression can be simplified as follows:

$$\frac{1}{2}\sqrt{(R\text{Sin}\theta)^2+\{R(1-\text{Cos}\theta)\}^2}=\frac{1}{2}\sqrt{R^2\text{Sin}^2\theta+R^2+R^2\text{Cos}^2\theta-2R^2\text{Cos}\theta}$$

$$=\frac{1}{2}\sqrt{2R^2(1-\text{Cos}\theta)}$$

$$=\frac{1}{2}\sqrt{2R^2\times 2\text{Sin}^2\left(\frac{\theta}{2}\right)}$$

$$=\frac{1}{2}\left\{2R\text{Sin}\left(\frac{\theta}{2}\right)\right\}$$

$$=R\text{Sin}\left(\frac{\theta}{2}\right)$$

Next, to find the approximate length of small arcs, the following formula is suggested.

$$\text{arc} \approx \sqrt{\frac{4}{3}(R\text{versine})^2 + (R\text{Sine})^2} \quad \text{for any small arc.}$$

If the arc is $s = R\theta$, then the above formula can be expressed in the following form:

$$s \approx \sqrt{\frac{4}{3}(R\text{versin}\,\theta)^2 + (R\text{Sin}\,\theta)^2} = \sqrt{\frac{4}{3}\{R(1-\text{Cos}\theta)\}^2 + (R\text{Sin}\theta)^2}$$

$$= \sqrt{\frac{4}{3}R^2\left\{2\text{Sin}^2\left(\frac{\theta}{2}\right)\right\}^2 + R^2\left\{2\text{Sin}\left(\frac{\theta}{2}\right)\text{Cos}\left(\frac{\theta}{2}\right)\right\}^2}$$

$$= \sqrt{\frac{4}{3}R^2\left\{4\text{Sin}^4\left(\frac{\theta}{2}\right)\right\} + R^2\left\{4\text{Sin}^2\left(\frac{\theta}{2}\right)\text{Cos}^2\left(\frac{\theta}{2}\right)\right\}}$$

$$= \sqrt{4R^2\text{Sin}^2\left(\frac{\theta}{2}\right)\left\{\frac{4}{3}\text{Sin}^2\left(\frac{\theta}{2}\right) + \text{Cos}^2\left(\frac{\theta}{2}\right)\right\}}$$

$$= \sqrt{4R^2\text{Sin}^2\left(\frac{\theta}{2}\right)\left\{1 + \frac{1}{3}\text{Sin}^2\left(\frac{\theta}{2}\right)\right\}}$$

$$= 2R\text{Sin}\left(\frac{\theta}{2}\right)\left\{1 + \frac{1}{3}\text{Sin}^2\left(\frac{\theta}{2}\right)\right\}^{\frac{1}{2}}$$

$$= 2R\text{Sin}\left(\frac{\theta}{2}\right)\left[1 + \frac{1}{2}\times\frac{1}{3}\text{Sin}^2\left(\frac{\theta}{2}\right) + \frac{\frac{1}{2}\times\left(\frac{1}{2}-1\right)}{1\times 2}\left\{\frac{1}{3}\text{Sin}^2\left(\frac{\theta}{2}\right)\right\}^2 + \dots\right]$$

$$= 2R\text{Sin}\left(\frac{\theta}{2}\right)\left\{1 + \frac{1}{6}\left\{\text{Sin}\left(\frac{\theta}{2}\right)\right\}^2 - \frac{1}{72}\left\{\text{Sin}\left(\frac{\theta}{2}\right)\right\}^4 + \dots\right\}$$

If θ is small, then $R\text{Sin}\,\theta$ is taken as the corresponding arc itself and so

$$R\text{Sin}\left(\frac{\theta}{2}\right) \approx R\left(\frac{\theta}{2}\right)$$

Hence, it follows that

$$\sqrt{\frac{4}{3}(R\text{Versin}\,\theta)^2 + (R\text{Sin}\,\theta)^2} \approx 2R\left(\frac{\theta}{2}\right)\left\{1+\frac{1}{6}\left(\frac{\theta}{2}\right)^2 - \frac{1}{72}\left(\frac{\theta}{2}\right)^4 + \ldots\right\}$$

$$\approx R\theta\left\{1+\frac{\theta^2}{24}-\frac{\theta^4}{1{,}152}+\ldots\right\}$$

Now, for sufficiently small values of θ such that θ^3 and higher powers of θ can be neglected, the series on the right side $\rightarrow R\,\theta$ or tends to the arc. To compute RSines and RCosines we begin by using Nilakantha's circumference and diameter ratio with the diameter and circumference of a circle taken as 113 and 355 units respectively. This gives

$$\pi = \frac{355}{113} = 3.14159292\ldots$$

Using this value, the value of R for computation of RSines is given by

$R = \dfrac{21600 \times 113}{2 \times 355} = 3437.746479 = 57^\circ 17' 44.8''$ or equal to 1 radian in modern terms.

Now the circumference is to be divided into a definite number of arcs of equal length for computing the RSines and RSine-differences at equal intervals. The RSine-differences and the corresponding RSines are interrelated and from the relation between them, the RSines can be derived successively using the recursive formula mentioned in the last three verses quoted earlier.

If the arc quadrant is divided into 24 equal parts, each part will be of angular measure $h = 225'$ and $24h = 90°$. The RSines of arcs of angular measures h, $2h$, $3h$, …, $24h$ are

$$J_1 = R\text{Sin}h,\ J_2 = R\text{Sin}2h,\ J_3 = R\text{Sin}3h,\ \ldots,\ J_{24} = R\text{Sin}24h$$

and the RSine-differences are

$$\Delta J_0 = J_1 - J_0 = J_1 \text{ where } J_0 = 0$$
$$\Delta J_1 = J_2 - J_1$$
$$\Delta J_2 = J_3 - J_2$$
$$\vdots$$
$$\Delta J_{23} = J_{24} - J_{23}$$

The last RSine is $J_{24} = R\text{Sin}24h = R$.
Now the penultimate RSine is J_{23} given by

$$J_{23} = R\text{Sin}23h = R\text{Sin}(24h - h) = R\text{Sin}(90° - h) = R\text{Cos}h$$
$$= \sqrt{R^2\text{Cos}^2 h} = \sqrt{R^2\left(1 - \text{Sin}^2 h\right)} = \sqrt{R^2 - R^2\text{Sin}^2 h} = \sqrt{R^2 - J_1^{\,2}}$$ where J_1

is the first RSine.

The last RSine-difference is given by

$$\Delta J_{23} = J_{24} - J_{23} = R - R\text{Sin}23h = R - R\text{Cos}h = R(1 - \text{Cos}h)$$

The difference between 23rd and 22nd RSines is $\Delta J_{22} = J_{23} - J_{22}$ given by the formula:

$$\Delta J_{22} = J_{23} \times \frac{2\Delta J_{23}}{R} + \Delta J_{23}$$ so that the 22nd RSine is $J_{22} = J_{23} - \Delta J_{22}$

$$= J_{23} - \left\{ J_{23} \times \frac{2\Delta J_{23}}{R} + \Delta J_{23} \right\}$$

The difference between 22nd and 21st RSines ($\Delta J_{21} = J_{22} - J_{21}$) can be obtained from the formula $\Delta J_{21} = J_{22} \times \frac{2\Delta J_{23}}{R} + \Delta J_{22}$ so that the 21st RSine is

$$J_{21} = J_{22} - \Delta J_{21}$$
$$= J_{22} - \left\{ J_{22} \times \frac{2\Delta J_{23}}{R} + \Delta J_{22} \right\}$$

Similarly, the difference between 21st and 20th RSines ($\Delta J_{20} = J_{21} - J_{20}$) can be obtained from the formula $\Delta J_{20} = J_{21} \times \frac{2\Delta J_{23}}{R} + \Delta J_{21}$ so that the 20th RSine can be calculated from

$$J_{20} = J_{21} - \Delta J_{20}$$
$$= J_{21} - \left\{ J_{21} \times \frac{2\Delta J_{23}}{R} + \Delta J_{21} \right\}$$

In general the difference between $(r+1)^{th}$ and r^{th} RSines $\Delta J_r = J_{r+1} - J_r$ is given by $\Delta J_r = J_{r+1} \times \frac{2\Delta J_{23}}{R} + \Delta J_{r+1}$ so that the rth RSine can be calculated from

$$J_r = J_{r+1} - \Delta J_r$$
$$= J_{r+1} - \left\{ J_{r+1} \times \frac{2\Delta J_{23}}{R} + \Delta J_{r+1} \right\}$$ where $r = 22, 21, 20, \ldots, 1, 0$ and

$$\Delta J_{23} = J_{24} - J_{23}$$
$$= R - R\text{Sin}23h$$
$$= R - R\text{Cos}h.$$

If we denote $2\left(\frac{\Delta J_{23}}{R}\right)$ by λ, then the above formulae reduce to the following form

$\Delta J_r = \lambda \times J_{r+1} + \Delta J_{r+1}$ and $J_r = J_{r+1} - \lambda \times J_{r+1} - \Delta J_{r+1}$ where $r =$ 22, 21, 20, …, 1, 0 and $\lambda = 2\left(\frac{\Delta J_{23}}{R}\right) = 2(1 - \text{Cos}h)$.

Again since $24h = 90°$; $23h = 90° - h$, $24h = 90° - 2h$, etc. and so

$$\Delta J_{22} = J_{23} \times \frac{2\Delta J_{23}}{R} + \Delta J_{23} \text{ which implies}$$

$$\Delta J_{22} = R\text{Sin}23h \times \frac{2R(1-\text{Cos}h)}{R} + R(1-\text{Cos}h)$$

i.e.; $\Delta J_{22} = R\text{Sin}\,(90° - h) \times 2\,(1 - \text{Cos}h) + R(1 - \text{Cos}h)$ and $J_{22} = J_{23} - \Delta J_{22}$

$$= R\text{Sin}\,(90° - h) - \{R\text{Sin}\,(90° - h) \times 2\,(1 - \text{Cos}h) + R(1 - \text{Cos}h)\}$$

Also, $\Delta J_{21} = J_{22} \times \frac{2\Delta J_{23}}{R} + \Delta J_{22}$ takes the form

$$\Delta J_{21} = J_{22} \times 2(1 - \text{Cos}h) + J_{23} - J_{22}$$

i.e.; $\Delta J_{21} = R\text{Sin}22h \times 2(1 - \text{Cos}h) + R\text{Sin}23h - R\text{Sin}22h$

$\Delta J_{21} = R\text{Sin}\,(90° - 2h) \times 2\,(1 - \text{Cos}h) + R\text{Sin}\,(90° - h) - R\text{Sin}\,(90° - 2h)$

and $J_{21} = J_{22} - \Delta J_{21}$ which is equal to

$R\text{Sin}\,(90° - 2h) - \{R\text{Sin}(90° - 2h) \times 2(1 - \text{Cos}h) + R\text{Sin}\,(90° - h) - R\text{Sin}\,(90° - 2h)\}$

Similarly,

$$\Delta J_{20} = R\text{Sin}\,(90^\circ-3h)\times 2(1-\text{Cos}h)+R\text{Sin}\,(90^\circ-2h)-R\text{Sin}\,(90^\circ-3h)$$

and

$$J_{20} = R\text{Sin}(90^\circ-3h)-\{R\text{Sin}(90^\circ-3h)\times 2(1-\text{Cos}h)+R\text{Sin}(90^\circ-2h) - R\text{Sin}(90^\circ-3h)\}$$

In general,

$$\Delta J_{24-r} = R\text{Sin}[90^\circ-(r-1)h]\times 2(1-\text{Cos}h) + R\text{Sin}[90^\circ-(r-2)h] - R\text{Sin}\,[90^\circ-(r-1)h]$$

$$J_{24-r} = R\text{Sin}(90^\circ-(r-1)h)-R\text{Sin}(90^\circ-(r-1)h)\times 2(1-\text{Cos}h) - R\text{Sin}(90^\circ-(r-2)h)+R\text{Sin}(90^\circ-(r-2)h)$$

Divide throughout by R and denote $90 - rh$ by B, these above two results are of the form

$$\Delta\text{Sin}B=\text{Sin}(B+h)\times 2(1-\text{Cos}h)+\text{Sin}(B+2h)-\text{Sin}(B+h) \text{ and}$$

$$\text{Sin}B=\text{Sin}(B+h)-\{\text{Sin}(B+h)\times 2(1-\text{Cos}h)+\text{Sin}(B+2h)-\text{Sin}(B+h)\}$$

Or, $\Delta\text{Sin}B=\text{Sin}(B+h)\times 2(1-\text{Cos}h)+\text{Sin}(B+2h)-\text{Sin}(B+h)$ and

$$\text{Sin}B=\text{Sin}(B+h)-2\text{Sin}(B+h)+2\text{Sin}(B+h)\text{Cos}h-\text{Sin}(B+2h)+\text{Sin}(B+h)$$

Or, $\Delta\text{Sin}B=\text{Sin}(B+h)\times 2(1-\text{Cos}h)+\text{Sin}(B+2h)-\text{Sin}(B+h)$

And, $\text{Sin}B=2\text{Sin}(B+h)\text{Cos}h-\text{Sin}(B+2h)$

From the second we get

$$\text{Sin}(B+2h)=2\text{Sin}(B+h)\text{Cos}h-\text{Sin}B$$

Replacing $B+2h$ by A, the result reduces to the form

$$\text{Sin}A=2\text{Sin}(A-h)\text{Cos}h-\text{Sin}(A-2h)$$

which is true since the right side of the foregoing can be expressed in the form

$$\text{Sin}(A-h+h)+\text{Sin}(A-h-h)-\text{Sin}(A-2h) = \text{Sin}A+\text{Sin}(A-2h)-\text{Sin}(A-2h) = \text{Sin}A$$

thus validating the sine formula given by Nilakantha.

Also, since $\Delta\text{Sin}B = \text{Sin}(B+h) - \text{Sin}B$ and

$$\begin{aligned} &\text{Sin}(B+h)\times 2(1-\text{Cos}h) + \text{Sin}(B+2h) - \text{Sin}(B+h) \\ &= 2\text{Sin}(B+h) - 2\text{Cos}h\times\text{Sin}(B+h) + \text{Sin}(B+2h) - \text{Sin}(B+h) \\ &= \text{Sin}(B+h) - \{\text{Sin}(B+h+h) + \text{Sin}(B+h-h)\} + \text{Sin}(B+2h) \\ &= \text{Sin}(B+h) - \{\text{Sin}(B+2h) + \text{Sin}B\} + \text{Sin}(B+2h) \\ &= \text{Sin}(B+h) - \text{Sin}B = \Delta\text{Sin}B. \end{aligned}$$

This verifies Nilakantha's formula for computating RSine-differences. Thus from the early beginnings in Aryabhata's work for a thousand years, Indian trigonometry reached a high level of sophistication in analysis and computation, while remaining true to its Aryabhatan roots.

Notes

1. The value of radius (R) = 3438 probably predates Aryabhata, but his is the first surviving text to record it.
2. An interesting application of this relationship is in the calculation of the position of the earth. The length of the arc from winter solstice to summer solstice is approximately 176 degree and 18 minutes. Assuming that crd 176°18′ = 6872 minutes, it follows that half the chord is 3436′. Using the Pythagorean theorem, the distance from the centre of the circle to this chord is $\sqrt{(3438^2 - 3436^2)} = 117'$. In other words, this chord is 117 minutes, almost two full degrees off-centre. This arises from the fact that the lengths of the four seasons are not equal in length. Winter solstice to spring equinox is the shortest, 89 days long. Spring equinox to summer solstice is almost 90 days, with the summer solstice to autumnal equinox being the longest, just over 93 days, and the autumnal equinox to the winter solstice being almost 93 days. Since the sun moves at a constant speed, the unequal lengths of the various seasons would indicate that the earth is off-centre. It was Hipparchus who first tackled the problem of calculating the position of the earth.
3. This is a slight modification of the translation in Mallayya (2009: 5).
4. A *rasi* means literally a 'heap' or 'quantity', but in a more technical context it represents a 'zodical sign' or 30 degrees.
5. There are earlier instances of the use of $\sqrt{10}$ as the preferred ratio. In Jaina texts *Anuyoga Dwara Sutra* and the *Triloko Sutra* from around the beginning of the first millennium AD, the circumference of the *Jambo* island (a cosmographic representation of the earth) is 316227 *yojanna*, 3 *krosa*, 122 *danda* and $13^1/_2$ *angula* where 1 *yojanna* is about 10 kilometres, 4 *krosa* = 1 *yojanna*, 2,000 *danda* = 1 *krosa* and 96 *angula* (meaning literally a 'finger's breadth') = 1 *danda*. The result is consistent with taking

the circumference to be given by $\sqrt{10}d$ where the diameter (d) = 100,000 *yojanna*. The choice of the square root of 10 for π was convenient, since in Jaina cosmography islands and oceans always had diameters measured in powers of 10.

6. This is a slight modification of the translation in Mallayya (2009: 14–15).
7. This is a slight modification of the translation in Mallayya (2009: 14–15).
8. It can be shown that this rule is equivalent to Newton's interpolation formula up to first order term. Thus, arc = $nh + \theta$ and

$$RSin(nh+\theta) = d_1 + d_2 + d_3 + \ldots + d_n + \frac{\theta . d_{n+1}}{h}$$

$$= RSinh + (RSin2h - RSinh) + (RSin3h - Rsin2h) + \ldots$$

$$+ \left\{ RSinnh - RSin(n-1)h \right\} + \frac{\theta \{RSin(n+1)h - RSinnh\}}{h}$$

$$= RSinnh + \frac{\theta\{RSin(n+1)h - RSinnh\}}{h}$$

Denoting nh by x, this equation reduces to

$$RSin(x+\theta) = RSinx + \frac{\theta}{h}\{RSin(x+h) - RSinx\}.$$

i.e., $f(x+\theta) = f(x) + \frac{\theta}{h}\Delta f(x)$ where $f(x) = RSinx$ and

$$\Delta f(x) = f(x+h) - f(x).$$

If $\frac{\theta}{h} = u$ then $f(x + hu) = f(x) + u\Delta f(x)$.

This is exactly the Newton's interpolation formula up to first order term.

9. For further details, see Mallayya (2009: 25–27).
10. For details, see Mallayya (2004).
11. This is same as

$$d = \frac{1}{2}\left(d_g + d_b\right) - \frac{1}{2}\left(d_g - d_b\right)\frac{\theta}{h} \quad \text{for RSine and}$$

$$d = \frac{1}{2}\left(d_g + d_b\right) + \frac{1}{2}\left(d_g - d_b\right)\frac{\theta}{h} \quad \text{for RVersine}$$

where $h = 10° = 600'$.

12. This is a small text on spherical astronomy summarised in 56 verses in Sanskrit.
13. See *Golasara Siddhantadarpanam ca* of *Gargya Kerala Nilakantha* Ms No. T 846. B, Transcript copy by Paramesvara Sastry, C.1024.E (K.U.O.R.I and Mss Library, Trivandrum), iii vs.2.
14. For details see Mallayya (2004, 2009) from whose work this section of the chapter on Nilakantha's *Golasara* draws heavily.

6

Squaring the Circle: The Kerala Answer*

Introduction

Long back in the remote antiquity human beings must have become consciously aware of the existence of circles. Nature exhibited them in so many different ways: in the ripples of water at the drop of a stone, in the centre of a sunflower, in the shape of a sun or a full moon or even by looking into the eyes of others and seeing the pupils. However, for a long time, such observations did not lead to measurement. But once measurement occurred, there came the gradual realisation that there existed a relationship between circumference and diameter that would not change with the size of the circle. One of the earliest recorded attempts to measure the area of a circle is contained in the Ahmose Papyrus from ancient Egypt dating around 1650 BC. Problem 50 in that Papyrus states: 'A circular field has a diameter of 9 *chet* (or approximately 450 metres). What is its area?' The solution offered is in the form of a rule: 'Cut off one-ninth of the diameter and construct a square with the remainder. The area of the circle has the same area as the area of the square' (Joseph 2000). No explanation is offered for this rule although the ratio of the circumference to the diameter of this field (that is, π) implied by this

* This chapter is based on a paper entitled 'Kerala Mathematics: Motivation, Rationale and Method', which was presented by Dr Mallayya at an International Workshop in Kovalam, Kerala and published in an edited volume as Mallayya and Joseph (2009b).

calculation is 3.1605 which is more accurate than other ancients who seemed satisfied with 3.[1] Other attempts at greater accuracy included: *(a)* a Babylonian tablet from Susa dated 2000 BC containing an implicit value for π of $3\frac{1}{8}$ and *(b)* the *Sulbasutras* (800–500 BC) from India which, in attempting to construct a square equal to the area of a circle produced an implicit value for π of 3.088.[2]

Almost a thousand years after the Egyptian attempt, the Greek Hippocrates of Chios stated unambiguously that the areas of circles are in the ratio of the squares of their diameter which is equivalent mathematically to the statement that the area of a circle is a constant times the square of its radius. Incidentally, it was the same Hippocrates who raised the possibility of constructing with a ruler and compass a square equal in area to a circle. This was a precursor to the vain (but mathematically productive) attempts to 'square a circle' for about 2,300 years before it was recognised that such an exercise was impossible. But along the way, Euclid in the fourth century BC showed that the ratio of the circumference to the diameter lay between 3 and 4 (that is, $3 < \pi < 4$). And less than a hundred years later, Archimedes of Syracuse devised an ingenious solution by calculating the perimeters of the inscribed and circumscribed polygons with 12, 24, 48 and 96 sides within and around a circle, to obtain upper and lower bounds for π as 223/71 and 22/7 respectively. The upper bound 22/7 is now known to all schoolchildren and is often referred to in scholarly circles as the Archimedean value. Archimedes' method remained essentially unchanged, while better and better approximations to π were obtained by increasing the number of sides of the two polygons through a process of larger and larger doublings. In AD 499, as mentioned previously, Aryabhata obtained an implicit estimate to π as 3.14156 (correct to 4 decimal places), probably applying Archimedean method to polygons of 384 sides.[3] Even earlier, the Chinese Li Hui (c. AD 260) and Tsu Ching Chu (AD 480) arrived at accurate estimates using a slightly different method requiring only an inscribed regular polygon within a circle. The method involved calculating by successive applications of the Pythagorean theorem the perimeters of polygons of 12, 24, 48, ... sides to obtain the Aryabhatan value mentioned earlier in the case of Li and $3.1415926 < \pi < 3.1415927$ (i.e., correct to 7 decimal places) in the case of Tsu.[4] In 1429, the Persian mathematician Jamshid al-Kashi applied the Archimedes method to calculate the perimeter of a 3×2^{28} sided polygon and thereby obtained a value of π correct to 16 decimal places! No mathematician of that period or even 200 years later, using the Archimedean method, would achieve such a feat of accuracy.[5]

In 1593, Francios Viete became the first European mathematician to produce the following infinite sequence of mathematical operations to represent π. [6]

$$\frac{2}{\pi}=\sqrt{\frac{1}{2}\cdot\sqrt{\frac{1}{2}+\frac{1}{2}\cdot\sqrt{\frac{1}{2}+\frac{1}{2}\cdot\sqrt{\frac{1}{2}+\frac{1}{2}\cdots}}}}$$

This expression was hardly suitable for computing π, given the cumbersome numerical calculations involving the products of square root of 2 at each step in the calculation, although today we would use an iterative method. Further, the fact is the expression converges far too slowly so that it would take a large number of terms on the right-hand side to achieve any degree of accuracy for the estimate of π. It is interesting in this context that Viete used the Archimedean method rather than the aforementioned method to compute π. About 60 years later, John Wallis (1616–1703), the English mathematician, generally recognised as one of the most influential figures in the development of modern mathematics, gave π as an infinite product in his book *Arithmetica Infinitorium* (1655) which was subsequently converted into a form of a continued fraction by William Brouncker.[7] Neither of these expressions proved to be efficient in the computation of π and instead the European and Arab mathematicians continued to employ the Archimedean method in computing this ratio.

Almost 250 years earlier than Viete, the founder of the Kerala School, Madhava of Sangamagrama (1340–1425) is believed to have derived the equivalent of the following result.[8]

$$\frac{\pi}{4}=1-\frac{1}{3}+\frac{1}{5}-\frac{1}{7}+\ldots$$

It was soon realised that this series was not helpful in obtaining accurate estimates of the circumference for a given diameter (that is, estimating π) because of the slowness of the convergence of the series. This gave impetus to developments in two directions: *(a)* rational approximations by applying corrections (*samskara*) to partial sums of the series; *(b)* obtaining more rapidly converging series by transforming the original series. There was considerable work in both directions which are discussed in *Yuktibhasa* and *Kriyakramakari*. Appendix II of this chapter contains a discussion of these approximations.[9]

Motivation in the Kerala Work

In 1637, René Descartes wrote in *La Geometrie*: 'The ratios between straight and curved lines are not known, and I believe cannot be discovered by human minds, and therefore no conclusion based upon such ratios can be accepted as rigorous and exact' (Descartes 1954: 91). In the earliest written records containing Indian mathematics, the *Sulbasutras* (c. 500–800 BC), there are instructions for estimating the length of a curved line by laying a rope along the curve and then measuring the length of the rope.

As we saw in an earlier chapter, a thousand years later in Verse 10 of Aryabhata's *Aryabhatiya* there is provision for an accurate measure of the curved circumference of a circle in terms of its diameter (a straight line).[10] In the next verse there is an indication, somewhat cryptically, how a sine table may be derived by approximating small arcs by line segments. This idea of the existence of a mathematical relationship between a curved and straight line was the basic inspiration behind the work of the Kerala School of mathematics and astronomy.

A primary mathematical motivation for the Kerala work on infinite series was the recognition of the impossibility of arriving at an exact value for the circumference of a circle given the diameter (that is, what we would now refer to as the problem of the incommensurability of π).[11] Nīlakantha explained this conundrum in his *Aryabhatiyabhasya*:

> Why is only the approximate value (of circumference) given here? Let me explain. Because the real value cannot be obtained. If the diameter can be measured without a remainder, the circumference measured by the same unit (of measurement) will leave a remainder. Similarly, the unit which measures the circumference without a remainder will leave a remainder when used for measuring the diameter. Hence, the two measured by the same unit will never be without a remainder. Though we try very hard we can reduce the remainder to a small quantity but never achieve the state of 'remainderlessness'. This is the problem. (Unithiri and Anandhavardhan 2002: 149–153)

This explanation was prompted by Verse 10 in *Aryabhatiya* referred to in earlier chapter: 'Add 4 to 100, multiply by 8, and add 62,000. The result is approximately the circumference of a circle whose diameter is 20,000' (Shukla 1976)

The word 'approximately'[12] gave food for thought. Faced by this problem, the strategy followed by the Kerala mathematicians was succinctly stated in *Kriyakramakari*:

> Even by computing the results progressively, it is impossible theoretically to come to a final value. So, one has to stop computation at that stage of accuracy that one wants and take the final result arrived at ignoring previous results. (Sarma 1972c)

To understand what is being suggested here, it is necessary to remember that there have historically been two main approaches to calculating the circumference. The first approach, as mentioned earlier, that goes back to the time of Eudoxus (c. 375 BC) and Archimedes (c. 250 BC) if not earlier[13] is to inscribe and/or circumscribe the circle in regular polygons. The end result, in terms of modern mathematics, is a recursion relation involving square roots which, if repeated, gives increasingly accurate approximations for the circumference.[14]

The Kerala School adopted a second approach based on infinite series and integrals. Here, the circumference is obtained from devising a limiting procedure that allows the series to converge (however slowly) to the circumference as the number of terms grows.[15]

In the course of this demonstration, the following results were established:[16]

$$1 + 2 + 3 + \ldots + N = \frac{N(N+1)}{2} \quad (1)$$

$$1^2 + 2^2 + 3^2 + \ldots + N^2 = \frac{N(N+1)(2N+1)}{6} \quad (2)$$

$$1^3 + 2^3 + 3^3 + \ldots + N^3 = \frac{N^2(N+1)^2}{4} \quad (3)$$

It was then realised that for large N (i.e., small steps in the rectification of the circle) the following result applied:

$$1^k + 2^k + 3^k + \ldots + N^k \approx \frac{N^{k+1}}{k+1}$$

so that in the limit [for large N] we can replace

$$\sum_{n=1}^{N}\left(\frac{n}{N}\right)^{k} \approx \frac{N}{k+1}$$

However, the Kerala mathematicians did not have the calculus necessary for proceeding any further. Instead, they adopted a combination of geometry, algebra and intuition in their approach that involved finding the length of an arc by approximating it to a straight line. Known later as the method of direct rectification, it involves summation of very small arc segments and reducing the resulting sum to an integral. For the first time in the history of mathematics, the Kerala mathematician introduced and used the idea of asymptotic expansions in their derivations.[17]

The Kerala derivation is based on an interesting geometric technique. The tangent is divided up into *equal* segments while at the same time forcing a sub-division of the arc into *unequal* parts. This is required since the method involves the summation of a large number of very small arc segments, traditionally achieved by the first method, by the 'method of exhaustion',[18] where there was a sub-division of an arc into *equal* parts. The adoption of this 'infinite series' technique rather than the 'method of exhaustion' for implicitly calculating π was not through ignorance of the latter in Kerala mathematics. But, as Jyesthadeva points out in his *Yuktibhasa*, the former avoids tedious and time-consuming root-extractions at a time when mechanical aids to calculations were not available.[19]

There are other aspects of the 'tool kit' used by the Kerala mathematicians that need highlighting. The derivation of the arctan (and π) series employed two results in elementary mathematics which have a long history in India: *(a)* The 'Pythagorean' result[20] which dates back to the *Sulbasutras* and *(b)* the properties of similar triangles which is little more than the geometrical version of the 'Rule of Three' (*trairasika*), discussed in Chapter 3. However, it would seem that Verse 26 of the *ganita* section of *Aryabhatiya* is the more likely source of inspiration for the Kerala mathematicians.[21]

The Kerala derivation deploys an ingenious iterative re-substitution procedure[22] to obtain the binomial expansion for the expression $\frac{1}{1+x}$ and then proceeding through a number of repeated summations (*varamsamkalithas*) of series, arrive at what must be the most remarkable part of the derivation, an intuitive leap that leads to the following asymptotic formula:[23]

$$\lim_{n\to\infty} \frac{1}{n^{p+1}} \sum_{j=1}^{n} j^p = \frac{1}{p+1}$$

It was soon realised that the series

$$C = 4d - \frac{4d}{3} + \frac{4d}{5} - \frac{4d}{7} + \ldots$$

was not particularly useful for making accurate estimates of the circumference (C) given the diameter *(d)* because of the slowness of the convergence of the series.[24] This gave impetus to the applications of corrections to partial sums of the series and to the obtaining of the more rapidly converging series by transforming the original series.

It is clear that in deriving what we would know call the arctan (and the π) series, the Kerala School showed both an awareness of the principle of integration and an intuitive perception of small quantities and operations with such quantities. However, having come so near to the formulation of the crucial concept of the 'limit' of a function, they shied away from developing the methods and algorithms of calculus, being perfectly content with the geometrical approach which their European counterparts soon replaced with calculus.[25]

The Rationale and Method: A Case Study of the Madhava–Gregory Series[26]

In the fourteenth century AD, Madhava of Sangamagramma discovered the remarkable connection between circumference of a circle and the reciprocals of all whole odd numbers that exists. In modern notation, the result may be expressed as a converging infinite series of the form.

$$\frac{\pi}{4} = 1 - \frac{1}{3} + \frac{1}{5} - \frac{1}{7} + \ldots \qquad (*)$$

However, as mentioned earlier, the convergence of this series is very slow,[27] and it would require a large number of terms on the right-hand side of (*) to approach a reasonably accurate value of π. As mentioned earlier, the desire for speeding up the convergence motivated some of the most innovative work on the part of the Kerala mathematicians which is discussed in the Appendix II of this chapter. This series was also discovered in Europe by Gregory in 1671 and Leibniz in 1673.

In the derivation of (*), the Kerala mathematicians were dependent on a 'toolkit', containing the four main results listed here. Only the first two will be examined here. The last two, already referred to earlier chapters, are sufficiently well known to require any further elaboration:[28]

1. Summation of a geometric series
2. Establishing the behaviour of a certain quotient taken to its limit
3. Properties of similar triangles
4. The Pythagorean theorem

Summation of a Geometric Series

The demonstration of this summation, as outlined in the Nilakantha's *Aryabhatiyabhasya* and Jyesthadeva's *Yuktibhasa*, may be restated thus:

Let r be a number such that $-1 < r < 1$. Show that

$$1 + r + r^2 + r^3 + \dots = \frac{1}{1-r} \qquad \text{(Infinite Series)}$$

and

$$1 + r + r^2 + r^3 + \dots + r^k = \frac{1-r^{k+1}}{1-r} \qquad \text{(Finite Series)}$$

<u>Proof</u> (A Visual Demonstration for $r < 1$ only)

Figure 6.1

Summation of a Geometric Series: A Visual Demonstration

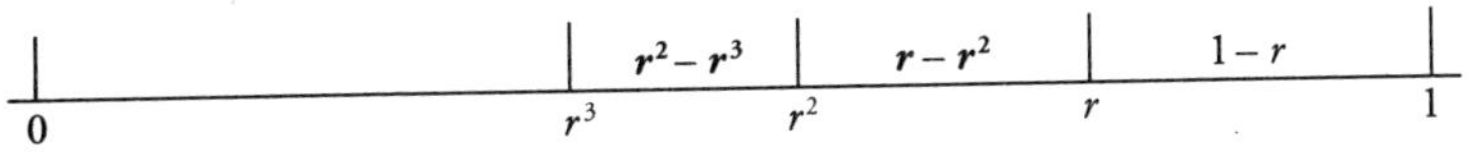

In Figure 6.1,

$$(1 - r) + (r - r^2) + (r^2 - r^3) + \dots = 1$$

Or,

$$(1 - r)\,(1 + r + r^2 + r^3 + \dots) = 1$$

Therefore,

$$1 + r + r^2 + r^3 + \ldots = \frac{1}{1-r}$$

Also implied in the development of this argument in the *Yuktibhasa*, is a more conventional geometrical demonstration.

In Figure 6.2, triangle *EFB* is similar to triangle *ABC* by scale factor *r*, and *AC* and *AF* are each of length 1. The other smaller triangles are similar with scale factors r^2, r^3, and so on. Since *ABC* and *DEC* are also similar, it would follow that

$$\frac{AB}{DE} = \frac{AC}{DC} \tag{1}$$

But $AC = 1$, $DE = 1$ and $DC = 1 - r$
Therefore (1) reduces to

$$AB = \frac{1}{DC} = \frac{1}{1-r}$$

But

$$AB = 1 + r + r^2 + r^3 + \ldots$$

So

$$\mathbf{1 + r + r^2 + r^3 + \ldots = \frac{1}{1-r}} \tag{2}$$

In the finite case,

$$(1 - r) + (r - r^2) + (r^2 - r^3) + \ldots + (r^k - r^{k+1}) = 1 - r^{k+1}$$

Figure 6.2
Summation of Geometric Series: A Geometric Demonstration

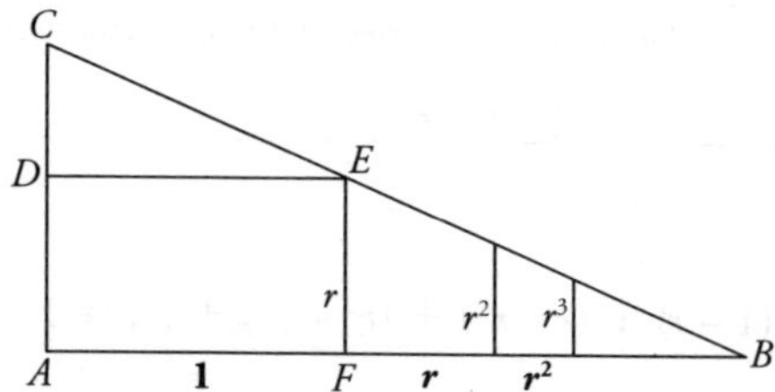

Factor out $(1 - r)$ from all terms on left-hand side and re-arrange to get

$$1 + r + r^2 + r^3 + \ldots + r^k = \frac{1-r^{k+1}}{1-r} \qquad (3)$$

Behaviour of a Quotient taken to its limit[29]

A Digression: It is an interesting point of comparison that in European mathematics this quotient arose from two sets of problems:

1. Finding the area under a curve
2. Construction of a tangent to a curve

(a) **Area under a curve $y = x^2$ (For $r < 1$)**

In Figure 6.3

$$\text{Area of (I)} = 1 \times (1 - r) = (1 - r)$$
$$\text{Area of (II)} = r^2 (r - r^2) = r^3 (1 - r)$$
$$\text{Area of (III)} = r^4 (r^2 - r^3) = r^6 (1 - r)$$
$$\text{Total area of (I)} + \text{(II)} + \text{(III)} = (1 - r)(1 + r^3 + r^6)$$

So the area of an endless staircase generated by increasing powers of r, as shown in Figure 6.3, is

$$(1 - r) + r^3(1 - r) + r^6(1 - r) + \ldots$$
$$= (1 - r)(1 + r^3 + r^6 + \ldots)$$
$$= (1 - r)\cdot\frac{1}{1-r^3} = \frac{1-r}{1-r^3}$$

Or more generally

$$=(1-r).\frac{1}{1-r^k} = \frac{1-r}{1-r^k}$$

Question: What happens to $\frac{1-r}{1-r^k}$ as r gets nearer to 1 and k increases?

If $k = 2$

$$\frac{1-r}{1-r^3} = (1 + r + r^2)^{-1} = 1/3 \text{ as } r \to 1$$

Figure 6.3

Area under a Curve

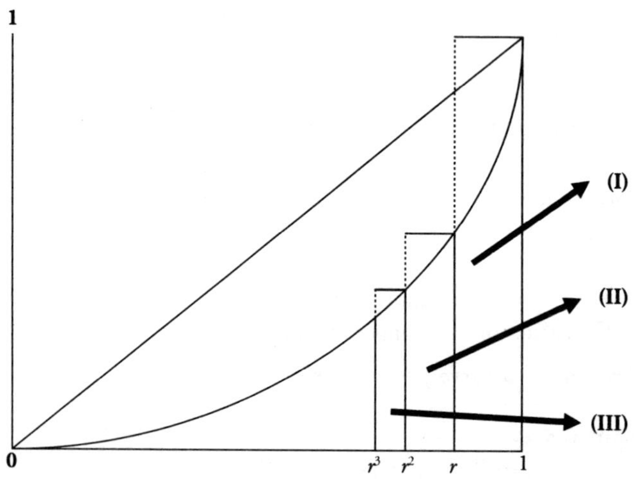

If $k = 3$

$$\frac{1-r}{1-r^4} = (1 + r + r^2 + r^3)^{-1} = \tfrac{1}{4} \text{ as } r \to 1$$

If $k = 4$

$$\frac{1-r}{1-r^5} = (1 + r + r^2 + r^3 + r^4)^{-1} = \tfrac{1}{5} \text{ as } r \to 1$$

Or more generally,

$$\frac{1-r}{1-r^{k+1}} = (1 + r + r^2 + r^3 + r^4 + \ldots + r^k)^{-1} = \frac{1}{k+1} \text{ as } r \to 1 \quad (5)$$

(b) Construction of a Tangent to a Curve

In Figures 6.4(a) and (b), the slope of the tangent line to the curve $y = x^2$ at P is

$$\frac{x^2-1}{x-1} = \frac{-(1-x^2)}{-(1-x)} = \frac{1-x^2}{1-x} = 1 + x = 2 \text{ as } x \to 1 \quad (6)$$

$$\tan\theta = \frac{\text{Rise}}{\text{Run}} = \frac{x^2-1}{x-1}$$

Figure 6.4

A Tangent to a Circle

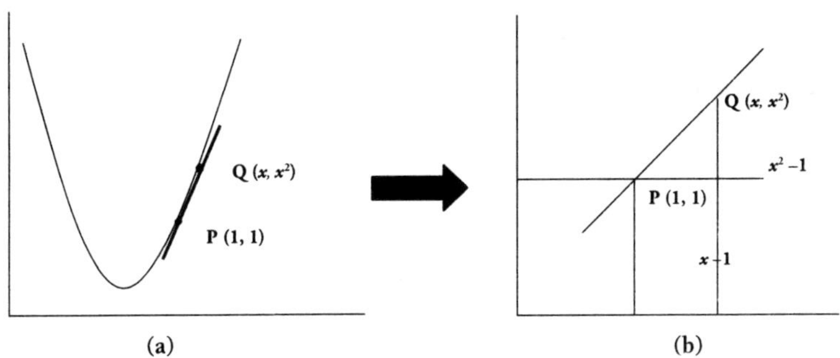

It is easily shown that for the curve $y = x^3$, the slope of the tangent to that curve at P is

$$\frac{1-x^3}{1-x} = 1 + x + x^2 = 3 \text{ as } x \to 1$$

Or more generally, for the curve $y = x^k$, the slope of the tangent to that curve at P is

$$\frac{1-x^{k+1}}{1-x} = 1 + x + x^2 + x^3 + x^4 + \cdots + x^k = k + 1 \text{ as } x \to 1 \qquad (7)$$

In Indian mathematics, the problem arose from an interest in series which has a long history probably starting from the Vedic times, with notable work by Jains followed by the synthesis of the result known by the Aryabhatan School. It arose, also somewhat unusually, from an interest peculiarly Indian.

<u>The Indian Dimension:</u> A Problem from *Sunyaganita* ('Operations with Zero'): What is Zero divided by Zero?

The question may be rephrased as

$$\text{Evaluate } \frac{x^2-1}{x-1} \text{ as } x \to 1 \qquad (8)$$

Note that when $x = 1$, (8) becomes 0/0 which is clearly not possible since division is a form of inverse multiplication so that

$$\frac{0}{0} = 0 \times \mathbf{?} = 0 \text{ where ? can be any number except } 0$$

So what is the solution to (8)?
Recall that:

$$\frac{1-x^{k+1}}{1-x} = 1 + x + x^2 + \ldots + x^k$$

For $k = 1$ and where $x \to 1$,

$$\frac{1-x^2}{1-x} = 1 + x = 2$$

For $k = 2$ and where $x \to 1$,

$$\frac{1-x^3}{1-x} = 1 + x + x^2 = 3$$

For the general case k where $x \to 1$,

$$\frac{1-x^{k+1}}{1-x} = 1 + x + \ldots + x^k = k + 1$$

The Madhava–Gregory Series: Derivation

Recall the infinite series given earlier which is reproduced here.

$$\frac{\pi}{4} = 1 - \frac{1}{3} + \frac{1}{5} - \frac{1}{7} + \ldots \qquad (*)$$

It is today a simple affair using calculus and basically involves <u>five</u> steps. In Figure 6.5, *Apq* is a sector of a circle of unit radius *OA*.

Figure 6.5

Derivation of the Madhava–Gregory Series

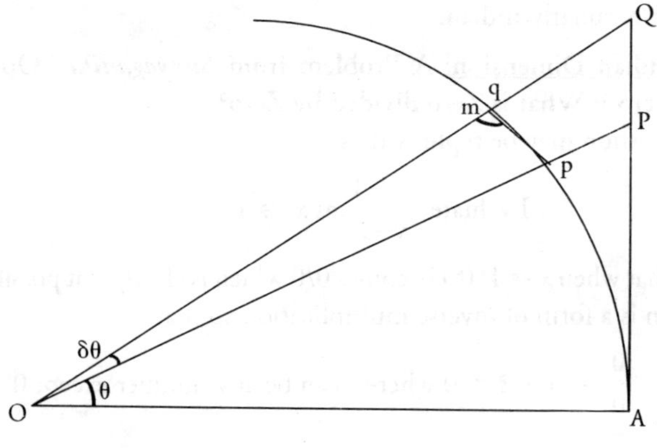

Step 1: $\partial\theta$ = arc $pq \approx pm = \dfrac{\partial(\tan\theta)}{(1+\tan^2\theta)}$ where $OA = Op = 1$

Step 2: So $\theta = \int \dfrac{dt}{1+t^2} = \int (1 - t^2 + t^4 - \ldots)dt = t - \dfrac{t^3}{3} + \dfrac{t^5}{5} - \cdots$ where $t = \tan\theta$

Step 3: It is next shown that the last term of the series in Step 2:

$$(-1)^{n+1}.\frac{t^{2n+2}}{(1+t^2)}dt \to 0 \text{ as } n \to \infty \text{ if } |t| \leq 1$$

Step 4: Or, arctanθ can be represented as an infinite series of the form:

$$\text{Arctan } \theta = \tan\theta - \frac{\tan^3\theta}{3} + \frac{\tan^5\theta}{5} - \ldots \text{ if } |\tan\theta| \leq 1 \qquad (*)$$

Step 5: For $\tan\theta = 1$ or $\theta = 45° = \pi/4$ radians, the above series (*) becomes:

$$\frac{\pi}{4} = 1 - \frac{1}{3} + \frac{1}{5} - \ldots \qquad (**)$$

(*) was first investigated in Europe by the Scottish mathematician, James Gregory, in 1671. Two years later, the German philosopher and mathematician, Gottfried Wilhelm Leibniz, discovered (**) using a different approach to that of Gregory.[30] But nearly three centuries earlier, both series were known and used in Kerala.

It is now proposed to show how the Kerala mathematicians derived the series given in (**).

Figure 6.6(a) represents a circle of unit radius (that is, $AC = CD = 1$) with the angle at the centre of the circle ($<ACB$) equalling 45 degrees. AB is the tangent at point A. It would follow that since $AB = 1$, circumference (C) = 2π and arc $AD = \frac{1}{8}C = \pi/4$. Figure 6.6(b) represents the line segment AB being divided into a number of sections (say ***n***) of *equal* length. Note that the length of each section is **1/*n***. Also, note that at the same time arc AD is cut into *n unequal* sections. The objective is to estimate the lengths of each of these small arcs and sum these estimates to obtain in turn an estimate of the length of arc AD which we know equals $\pi/4$.

Without loss of generality, consider the special case of $n = 5$, as shown in Figure 6.6(c). We wish to estimate the portion of the arc labelled EF that corresponds to the line segment 3/5 to 4/5, labelled GH. In terms of Figure 6.6(d), the objective is to estimate FI which will be a good approximation for the arc FE when GH is very small.

Figure 6.6

Steps in the Derivation of the Madhava–Gregory Series Derivation

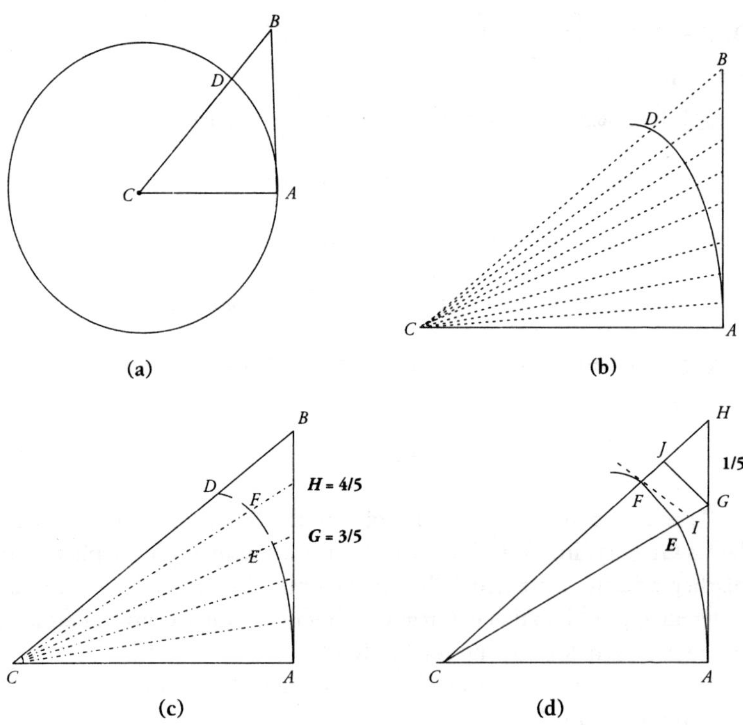

Proof: Right-angled triangles *GJH* and *CAH* are similar. So

$$\frac{GJ}{AC} = \frac{GH}{CH} \tag{1}$$

Since $AC = 1$ and $GH = \frac{1}{5}$, it would follow that:

$$GJ = \frac{1/5}{CH} \tag{2}$$

Also, since right-angled triangles *CFI* and *CJG* are similar, it would follow that

$$\frac{FI}{GJ} = \frac{CF}{CJ} \Rightarrow FI = \frac{GJ \cdot CF}{CJ}$$

But $CF = 1$. So

$$FI = \frac{GJ}{CJ} \tag{3}$$

When n is large, GH is small and so is JH. Therefore, CH is a good approximation to CJ

Or,

$$\frac{CH}{CJ} \approx 1$$

Using this fact, rewrite (3) as

$$FI = \frac{GJ}{CJ} \cdot \frac{CH}{CH}$$

So

$$FI = \frac{GJ}{CH} \tag{4}$$

Combining (4) with the earlier result given in (2), we get

$$FI = \frac{1/5 \big/ CH}{CH} = \frac{1/5}{CH^2} \tag{5}$$

Now, note that CH is the hypotenuse of right-angled triangle CAH. So,

$$CH^2 = CA^2 + AH^2 = 1 + (4/5)^2 \tag{6}$$

Combining (5) and (6) we get

$$FI = \frac{1/5}{1+(4/5)^2}$$

So an estimate of arc FE is

$$\text{arc } FE \approx FI = \frac{1/5}{1+(4/5)^2} \tag{7}$$

In the general case, the numerator (1/5) will be replaced by ($1/n$) and the (4/5) in the denominator of (7) will be replaced by a fraction that describes the length AH. In the present case of $n = 5$, the length of the arc *AD* is the sum of the estimates of five short arcs where each estimate will correspond to that calculated for arc *FE* in (7). In each case the numerator is 1/5 and the denominator is different, each going from $1 + (1/5)^2$ to $1 + (5/5)^2$.

The total estimate of arc length *AD* will be

$$\frac{1/5}{1+(1/5)^2}+\frac{1/5}{1+(2/5)^2}+\frac{1/5}{1+(3/5)^2}+\frac{1/5}{1+(4/5)^2}+\frac{1/5}{1+(5/5)^2} \tag{8}$$

What we are really interested is what happens to the summands as n increases from $n = 5$ to an arbitrary large number. (8) can then be generalised to

$$\frac{1/n}{1+(1/n)^2}+\frac{1/n}{1+(2/n)^2}+\frac{1/n}{1+(3/n)^2}+\ldots+\frac{1/n}{1+(n/n-1)^2}+\frac{1/n}{1+(n/n)^2} \tag{9}$$

Factoring out the numerator $1/n$ in (9) gives

$$\frac{1}{n}\left[\frac{1}{1+(1/n)^2}+\frac{1}{1+(2/n)^2}+\frac{1}{1+(3/n)^2}+\ldots+\frac{1}{1+(n-1/n)^2}+\frac{1}{1+(n/n)^2}\right] \tag{10}$$

Recall the earlier result for the summation of a geometric series

$$\frac{1}{1-r} = 1 + r + r^2 + r^3 + \ldots$$

Replacing r by $-s$ where $s > 0$

$$\frac{1}{1+s} = 1 - s + s^2 - s^3 + \ldots \tag{11}$$

Applying (11) to each term in the bracket in (10)

$$\frac{1}{1+\left(\frac{1}{n}\right)^2}=1-\left(\frac{1}{n}\right)^2+\left(\frac{1}{n}\right)^4-\left(\frac{1}{n}\right)^6+\ldots$$

$$\frac{1}{1+\left(\frac{2}{n}\right)^2}=1-\left(\frac{2}{n}\right)^2+\left(\frac{2}{n}\right)^4-\left(\frac{2}{n}\right)^6+\ldots$$

$$\frac{1}{1+\left(\frac{3}{n}\right)^2}=1-\left(\frac{3}{n}\right)^2+\left(\frac{3}{n}\right)^4-\left(\frac{3}{n}\right)^6+\ldots$$

$$\frac{1}{1+\left(\frac{n-1}{n}\right)^2}=1-\left(\frac{n-1}{n}\right)^2+\left(\frac{n-1}{n}\right)^4-\left(\frac{n-1}{n}\right)^6+\ldots$$

The last term in the brackets in (9) reduces to

$$\frac{1}{n}\left[\frac{1}{1+\left(\frac{n}{n}\right)^2}\right]=\frac{1}{2n}$$ which tends to 0 as n increases.

Adding the foregoing set of equations column wise:
The first column sum of $(n-1)$ 1's gives $n-1$.
The second column sum, after factoring out the denominator n^2, gives:

$$-\frac{1}{n^2}\left[\cdot 1^2+2^2+3^2+\ldots+(n-1)^2\right]$$

Similarly, the subsequent column sums after factoring out of the denominator n^2 can be written as

$$\frac{1}{n^4}\left[1^4+2^4+3^4+\ldots+(n-1)^4\right]$$

$$-\frac{1}{n^6}\left[1^6+2^6+3^6+\ldots+(n-1)^6\right]$$

$$\vdots$$

Substituting the column sums given above in (10) gives

$$\frac{n-1}{n}-\frac{1}{n^3}\left[1^2+2^2+3^2+\ldots+(n-1)^2\right]+\frac{1}{n^5}\left[1^4+2^4+3^4+\ldots+(n-1)^4\right]-\frac{1}{n^6}\left[1^6+2^6+3^6+\ldots+(n-1)^6\right]+\ldots \tag{12}$$

From an earlier result regarding the behaviour of a quotient taken to the limit, we established that as n approaches infinity

$$\frac{1}{n^{k+1}}\left[1^k+2^k+3^k+\ldots+(n-1)^k\right]=\frac{1}{k+1} \tag{13}$$

Applying (13) to (12) gives

$$-\frac{1}{n^3}\left[1^2+2^2+3^2+\ldots+(n-1)^2\right]=-\frac{1}{3}$$

$$\frac{1}{n^5}\left[1^4 + 2^4 + 3^4 + \ldots + (n-1)^4\right] = \frac{1}{5} \tag{14}$$

$$-\frac{1}{n^7}\left[1^6 + 2^6 + 3^6 + \ldots + (n-1)^6\right] = -\frac{1}{7}$$

$$\vdots \qquad\qquad\qquad\qquad \vdots$$

Therefore, the length of arc AD or $\frac{\pi}{4}$ is given by the infinite series

$$\frac{\pi}{4} = 1 - \frac{1}{3} + \frac{1}{5} - \frac{1}{7} + \ldots \tag{**}$$

Note that this is equivalent to

$$C = 4d - \frac{4d}{3} + \frac{4d}{5} - \frac{4d}{7} + \ldots$$

which in turn was what *Kriyakramakari* reports Madhava as saying that the way to measure the circumference of a circle is to: 'Multiply the diameter by 4. Subtract from it and add to it alternately the quotients obtained by dividing four times the diameter by the odd integers 3, 5, ... etc' (Sarma 1972c).

Conclusion

The major breakthrough in Kerala mathematics was the appearance of mathematical analysis in the form of infinite series and their finite approximations relating to circular and trigonometric functions. The primary motivation for this work was a mixture of intellectual curiosity and a requirement for greater accuracy in astronomical computations. Demonstrations of these results may not be completely rigorous by today's standards, but they are nonetheless correct. And these demonstrations may well be chosen for a modern mathematics classroom because the approach is more intuitive and therefore more convincing.[31]

Appendix I

The *Yuktibhasa*: An Important Milestone*

The *Yuktibhasa* is believed by many to be the seminal expository text of the Kerala School of mathematics and astronomy. The name of the author is not mentioned anywhere in the text. Evidence (Sarma 2008: xxxiv–xxxvii) exists from other sources to indicate that the author was Jyesthadeva who lived during the period AD 1500–1610. What we know is that he was a younger contemporary of Nilakantha and taught by the same teacher, Damodara (son of Paramesvara who was in turn a student of the great Madhava himself). Jyesthadeva taught Acyuta Pisarati (1550–1621) who refers to his aged teacher in reverential terms in his work on the computation of eclipses *Uparaga Kriyakrama*. The other personal information that we have of the author of the *Yuktibhasa* is that he came from the Parannottu family whose house was located in the vicinity of the Alathiyur village, the place from which other notable figures of the Kerala School such as Paramesvara, Nilakantha and Acyuta Pisarati originated.

The *Yuktibhasa* was composed around 1530 as an elaboration and explanation of the rationale underlying the theories and computations found in Nilakantha's *Tantrasangraha*.[32] Although the whole text reads as one piece, without any divisions or sections, it treats a wide range of

*A recent translation of the *Yuktibhasa* into English by Professor K.V. Sarma (with explanatory notes by K. Ramasubramanian, M.D. Srnivas and M.S. Sriram), posthumously published in 2008 forms an important basis of the discussion of the methods of the Kerala School contained in this and other chapters of this book. I was very fortunate in being entrusted by the author with a copy of this translation long before it was published. I wish to pay a special tribute to the Professor Sarma whose long and skilful translation and interpretation of the seminal texts of Kerala mathematics and astronomy have made this remarkable episode in the history of mathematics more accessible and comprehensible to a wider public.

subjects in mathematics and astronomy. The conventional view is to classify this work in two parts, mathematics and astronomy respectively. The mathematical part is in turn sub-divided into seven chapters which are as follows:

1. The mathematical operations: addition and subtraction, different methods of multiplication, division, operations of squaring and square roots.
2. Ten algebraic problems and their solutions.
3. Fractions dealing with different arithmetical operations.
4. The Rule of Three, including the general and inverse rules.
5. Importance of the Rule of Three in mathematical and astronomical computations illustrated by the method for computating the current *Kali* day.[33]
6. Methods of determining the circumference of a circle of a given diameter. These include work on infinite series which predate discoveries made by Gregory, Leibniz.
7. Methods of derivation of the Indian Sines, including a detailed examination of the rationale underlying the different procedures. These procedures predate the work done by Newton.

The last two chapters contain the most innovative section of the book. They provide the subject matter for the discussion of the methodology of Kerala mathematics covered in the present book. Chapters 8 to 15 constitute Part II of the *Yuktibhasa.* These include a variety of topics such as planetary motion, procedures involving mean planets, true sun, true planets (Chapter 8); spheres of the earth and heavenly bodies (Chapter 9); atmosphere and asterisms corrections for the moon (Chapter 11); eclipse (Chapter 12); and moon's cusps and phases of the moon (Chapter 15). The treatment is extensive and constitutes almost 60 per cent of the total coverage of the *Yuktibhasa* (Sarma 2008, vol. II).

It is worth noting in conclusion that the term *yukti* (as in *Yuktibhasa*) has different connotations, with the common underlying idea of reaching a conclusion or drawing an inference. It describes a process or methodology of reasoning particularly pertinent in Indian scientific discourse. Nilakantha describes his great inspiration, Aryabhata, as *yukti-nidhi*, a treasure trove of all yuktis. Jystedhava's *Yuktibhasa* is a model of what good *yukti*s should be.

Appendix II

Rational Approximations of Circumference (and π): The Kerala Contribution

Introduction

In applying an infinite series approach to estimate the circumference, the Kerala mathematicians came across a serious difficulty. The problem is that the series relating to the circumference converges far too slowly. For example, summing the first 19 terms on the right-hand side of the circumference formula [$C(n)$], shown below and derived earlier as $C(n) = 4d\left[1-\frac{1}{3}+\frac{1}{5}-\dots\right]$ only gives the highly inaccurate estimate of π as 3.194.

The problem was tackled by the Kerala School in two directions: *(a)* deriving rational approximations by applying corrections to partial sums of the series; and *(b)* obtaining more rapidly converging series by transforming the original series. There was considerable work in both directions as shown in both *Yuktibhasa* and *Kriyakramakari*. The discussion that follows is based on these two texts.

Approximations Through Partial Corrections

In *Tantrasangraha*, Nilakantha gives a correction to the circumference formula [$C(n)$] which he credits to Madhava. Expressed in modern notation

$$C(n) = 4\left[1-\frac{d}{3}+\frac{d}{5}-\dots+(-1)^{n-1}\frac{d}{2n-1}+(-1)^{n}\,d\cdot F(n)\right] \qquad (1)$$

where d is the diameter of the circle, n the number of terms on the right-hand side and the last term is the correction. Note that without the correction and if $d = 1$, then (1) becomes the Madhava–Leibniz series discussed earlier.

Three types of corrections are considered.[34]

$$F_a(n) = \frac{1}{4n} \tag{2a}$$

$$F_b(n) = \frac{n}{(4n^2 + 1)} \tag{2b}$$

$$F_c(n) = \frac{n^2 + 1}{(4n^3 + 5n)} \tag{2c}$$

No detailed explanation is given as to how these corrections were arrived at. However, in both *Yuktibhasa* and *Kriyakramakari,* attempts are made at providing a rationale for the incorporation of corrections (2a) and (2b), with an indication that the methods were known to the *acharya* (teacher) who could be either Madhava or Nilakantha.

Rewrite (1) as

$$C(n) = 4d\left[1 - \frac{1}{3} + \frac{1}{5} - \ldots + (-1)^{n-1}\frac{1}{2n-1} + (-1)^n \cdot F(n)\right] \tag{2}$$

Replace n by $(n + 1)$ to form $C(n + 1)$ and subtract $C(n)$ from $C(n + 1)$. For the correction $F(n)$ to be 'ideal', $C(n)$ must be equal to $C(n + 1)$ when n is large, that is,

$$F(n) + F(n + 1) - \frac{1}{m} = 0 \qquad \text{where } m = 2n + 1 \tag{3}$$

The problem is therefore to find a function $F(n)$ which satisfies (3). Obviously, the condition is satisfied if $F(n) = F(n + 1) = \frac{1}{2m}$. But as the authors of *Kriyakramakari* point out, this is not a realistic assumption. So the problem, in terms of modern exposition, is to minimise the expression.

$$F(n) + F(n + 1) - \frac{1}{m} \text{ for large } n$$

By a process of trial and error, the authors then proceed to consider different values of $F(n)$ which would accomplish this purpose. If

$$F_a(n) = \frac{1}{4n} = \frac{1}{2m - 2},$$

Then

$$F_a(n+1) = \frac{1}{2m+2},$$

So that

$$F_a(n) + F_a(n+1) = \frac{1}{2m-2} + \frac{1}{2m+2} = \frac{m}{m^2-1}$$

Or

$$\frac{m}{m^2-1} - \frac{1}{m} = \frac{1}{m^3-m} = \frac{1}{4n(n+1)(2n+1)}$$

is the 'error' (*sthaulya*) that remains from the use of the correction term

$$F_a(n) = \frac{1}{4n} \text{ given as (2a) above}$$

For large n, the incorporation of this correction only gives a small numerical error, that is,

$$\lim \frac{1}{4n(n+1)(2n+1)} \Rightarrow 0 \; as \; n \Rightarrow \infty$$

The authors of *Kriyakramakari* added, '[The *acharya*] was not satisfied with the [size] of this error and tried to find out [more accurate] corrections' (Sarma 1972: 388).

A series of adjustments was made to the denominator of $F_a(n)$ with varying degrees of success until 'the *acharya* added four divided by itself'. One can only assume that this meant an addition of $\frac{4}{4n}$ to the denominator of $F_a(n)$.

$$F_b(n) = \frac{1}{4n + \frac{4}{4n}} = \frac{n}{4n^2+1} \qquad \text{given as (2b) above}$$

The authors of *Kriyakramakari* then proceed to calculate the error (*sthaulya*) in the case of the correction $F_b(n)$. Using a similar approach to the one used in calculating the 'error' in the case of $F_a(n)$ and assuming that $m = 2n + 1$, the error (E) in $F_b(n)$ can be shown to be[35]

$$E = \frac{4}{m^5+4m} = \frac{4}{(2n+1)^5 + 4(2n+1)}$$

As n increases, $E \rightarrow 0$ faster than the error of $Fa(n)$.

No explanation is given in any of the texts of Kerala mathematics as to how the last of the corrections [$F_c(n)$] is derived. However, Hayashi et al. (1990) offer an explanation of how all corrections were derived by the *acharya* (taken to be Madhava) which is both ingenious and convincing. This is based on two premises:

1. Madhava used the value of 355/113 for the ratio of circumference to diameter (π). This ratio was known in India from at least the ninth century AD.
2. Madhava used the so-called Euclidean algorithm of division known in India from the time of Aryabhata in the solution of indeterminate equations of the first degree.

Let
$$S(n) = 1 - \frac{1}{3} + \frac{1}{5} - \ldots + (-1)^{n-1}\frac{1}{2n-1} \tag{4a}$$
From (1)
$$\frac{\pi}{4} = 1 - \frac{1}{3} + \frac{1}{5} - \ldots + (-1)^{n-1}\frac{1}{2n-1} \quad + (-1)^n F(n) \tag{4b}$$

Therefore,

$$F(n) = |\, S(n) - \pi/4 \,| \tag{4c}$$

Take $\pi = 355/113$ and calculate from (4c) the values of $F(n)$ for $n = 1, 2, 3, 4, 5, \ldots$ to get the series of fractions

97/452, 161/1356, 551/6780, 2923/47460, 21153/427140, …

Finding no mathematical pattern in this series, the conjecture is that Madhava proceeded with the help of Euclidean division to calculate the nearest unit fraction of each term and obtain an approximate value.

Thus,

$F(1) = 97/452 = 1/(4 + 64/97) \approx 1/4$ or $1/5$
$F(2) = 161/1356 = 1/(8 + 68/161) \approx 1/8$
$F(3) = 551/6780 = 1/(12 + 168/551) \approx 1/12$
$F(4) = 2923/47460 = 1/(16 + 692/2923) \approx 1/16$
$F(5) = 21153/427140 = 1/(20 + 4080/21153) \approx 1/20$

Or $F(n) = \ldots = 1/(4n + a/b) \approx 1/4n$ is the first correction given as $F_a(n)$ in 2(a)

A search was then made for a rule for the fractions left out in the above procedure:

Thus,

$$\begin{aligned} 64/97 &= 1/(1 + 33/64) \approx 1/1 \text{ or } 1/2 \\ 68/161 &= 1/\,(2 + 25/68) \approx 1/2 \\ 168/551 &= 1/(3 + 47/168) \approx 1/3 \\ 692/2923 &= 1/(4 + 155/692) \approx 1/4 \\ 4080/21153 &= 1/(5 + 753/4080) \approx 1/5 \end{aligned}$$

....

....

$$= 1/(n + c/d) \approx 1/n$$

Therefore, the second correction is

$$F_b(n) = \frac{1}{4n+\frac{1}{n}} = \frac{n}{4n^2+1} \text{ as given in 2(b)}$$

The procedure continues by again finding a rule to incorporate the fractions left out in the above calculations.
Thus,

$$\begin{aligned} 33/64 &= 1/(1 + 31/33) \approx 1/1 \text{ or } 1/2 \\ 25/68 &= 1/(2 + 18/25) \approx 1/2 \\ 47/168 &= 1/(3 + 27/47) \approx 1/3 \\ 155/692 &= 1/(4 + 72/155) \approx 1/4 \\ 753/4080 &= 1/(5 + 315/753) \approx 1/5 \end{aligned}$$

....

....

$$= 1/(n + e/f) \approx 1/n$$

Therefore, the third correction 2(c) is

$$F_c(n) = \frac{1}{4n+\frac{1}{n+\frac{1}{n}}} = \frac{n^2+1}{4n^3+5n}$$

It is possible that a fourth approximation was attempted but finding no pattern in the fractions generated, this procedure was aborted.

Rapidly Converging Series For Circumference

Another approach to obtaining accurate estimates of the circumference is to derive more rapidly converging series. The *Yuktibhasa* contains a number of such series either obtained from the transformation of the original series or by extending the theory of the 'errors' discussed in the last section.[36]

Grouping the terms of the series in

$$C = 4d\,[1 - 1/3 + 1/5 - 1/7 + 1/9 - 1/11 + \ldots] \qquad (5)$$

two by two gives

$$C = 4d\,[(1 - 1/3) + (1/5 - 1/7) + (1/9 - 1/11) + \ldots]$$

Hence,

$$C = 8d\,[1/(2^2 - 1) + 1/(6^2 - 1) + 1/(10^2 - 1) + \ldots] \qquad (6)$$

Regrouping (5) two by two but leaving out the first term gives

$$C = 4d\,[1 - (1/3 - 1/5) - (1/7 - 1/9) - (1/11 - 1/13) - \ldots]$$

Hence,

$$C = 4d - 8d/(4^2 - 1) - 8d/(8^2 - 1) - 8d/(12^2 - 1) - \ldots \qquad (7)$$

Taking half the sum of (6) and (7) and incorporating the last term gives

$$C(n) = 2d + \frac{4d}{2^2 - 1} - \frac{4d}{4^2 - 1} + \ldots + (-1)^n \frac{4d}{(2n+1)^2 + 2} \qquad (8)$$

There were other formulae which are developed along similar lines as the derivation of 'errors' discussed in the last section. These include[37]

$$C = \sqrt{12} \times \left(1 - \frac{1}{3 \cdot 3} + \frac{1}{3^2 \cdot 5} - \frac{1}{3^3 \cdot 7} + \ldots\right) \qquad (9)$$

$$C = 3d + \left(\frac{4d}{3^3 - 3} - \frac{4d}{5^3 - 5} - \frac{4d}{7^3 - 7} + \ldots\right) \qquad (10)$$

Note that if $d = 1$, (10) will become

$$C = 3 + 4\left(\frac{1}{2\cdot3\cdot4} - \frac{1}{4\cdot5\cdot6} + \frac{1}{6\cdot7\cdot8} + \ldots\right) \quad (11)$$

Conclusion

To illustrate the efficacy of the first approach (that is, achieving greater accuracy through incorporation of partial corrections), consider the incorporation of the third correction term $F_c(n)$ to the circumference series ($C(n)$) introduced at the beginning of this Appendix and where $n = 11$, the implicit estimate of π is 3.1415926529 which is correct to 8 places. Interest in improving the accuracy of the estimate continued for a long time, so that as late as the nineteenth century the author of *Sadratnamala* estimated the circumference of a circle of diameter 10^{18} as: 314,159,265,358,979,324 correct to 17 places![38]

It is clear from the foregoing discussion that interest in the application of rational approximations to infinite series had taken the Kerala mathematicians a long way. They showed some understanding of convergence, of the notion of rapidity of convergence and an awareness that convergence can be speeded up by transformations, several of which were worked out with considerable dexterity. And their interest in increasing the accuracy of their estimates continued for a long time, as late as the beginning of the nineteenth century.

Notes

1. For further details on the Egyptain attempt, see Joseph (2000: 82–84). It is interesting in this context that in both the Bible and the Talmud the implicit value for π is 3. This is not surprising given that the mathematically sophisticated Babylonians calculated the area of a circle as three times the square of the radius which implied a value of 3 for π.
2. For further details of the *Sulbasutra* method, see Joseph (2000: 232–33).
3. It is interesting to note that Brahmagupta (b. AD 598) was content with an implicit value for π equal to square root of 10. This was possibly the influence of Jaina cosmography in which islands and oceans had diameters always measured in powers of 10 or due to an unwarranted inference that since perimeters of polygons of 12,

24, 48 and 96 sides inscribed in a circle of diameter 10 are given by $\sqrt{965}$, $\sqrt{981}$, $\sqrt{986}$ and $\sqrt{987}$ respectively, a further doubling of sides would lead to the perimeter approaching 1,000 so that $\pi = \sqrt{1{,}000}/10 = \sqrt{10}$.

4. For further details, see Joseph (2000: 193–95).
5. For example, Leonardo of Pisa (1170–1250), using a 96-sided polygon obtained the value 864/275 = 3.141818 (correct to 3 decimal places) and the German astronomer Rheticus (1514–1576) in the course of constructing sine tables obtained an implicit value for π correct to 8 decimal places if we credit him with the knowledge of trigonometry as it exists today.
6. This is contained in his work entitled *Variorum De Rebus Mathematics Responsorum Liber VIII.*
7. Wallis's attempt to express π as an infinite product may be expressed as

$$\frac{4}{\pi} = \frac{3\times3\times5\times5\times7\times7\times\ldots}{2\times4\times4\times6\times6\times8\times\ldots}$$

In 1656, Brouncker converted Wallis's result into an expression involving continued fractions

$$\frac{4}{\pi} = 1 + \cfrac{1^2}{2 + \cfrac{3^2}{2 + \cfrac{5^2}{2 + \cfrac{7^2}{2 + \cfrac{9^2}{\vdots}}}}}$$

8. In *Yuktidipika*, a sixteenth-century commentary on Nilakantha's *Tantrasangraha* by Sankara Varier, Madhava is quoted as saying that the way to measure the circumference of a circle is to 'Multiply the diameter by 4. Subtract from it and add to it alternately the quotients obtained by dividing four times the diameter to the odd integers 3, 5 etc.' In other words, if d is the diameter of the circle,

$$C = 4d - \frac{4d}{3} + \frac{4d}{5} - \frac{4d}{7} + \ldots$$

which is equivalent to (**). However, at the end of each chapter, Sankara states that his exposition of the subject is based on Jyesthadeva's *Yuktibhasa.* It may be remembered that Jyesthadeva was a younger contemporary and student of Nilakantha.

9. The discussion is based on the *Yuktibhasa* and *Kriyakramakari* and contains application of some ingenious corrections to partial sums of the series as well as derivation of more rapidly converging series, both with the purpose of obtaining accurate estimates of the circumference for given diameters.
10. It is interesting to note in this context that the Indians from the very beginning took the attitude that the radial distance should be measured in the same units in which the length of the circumference is measured. In fact, Chapter 6 of the *Yuktibhasa* is introduced with the statement: 'Now is the stated the method to know the measure of the circumference of a circle in terms of its diameter which forms the side of a

square, the said side being taken to of measure unity in some unit like the cubit or *angula*' (Sarma 2008: 44). This would have been consistent with the modern concept of radians had they not retained the Babylonian sexagesimal division of a circle into 360 parts.

11. Note that approximate formulae for π are good enough for practical purposes. Archimedes of Syracuse (287–212 BC) obtained the value $223/71 < \pi < 22/7$ by considering a regular polygon of 96 sides to 2 decimal places. The value for π of 22/7 is perfectly acceptable even today among engineers.
12. Some Western historians of mathematics [notably Kaye (1908) and Kline (1972)] have argued that the Indians were not aware of the fact that π could never be exactly determined. Their confusion may have arisen because of the early mistranslation of the word *asana* as 'approximate' or 'rough value' as in the quotation given. The word *asana* is more subtle since it conveys the notion of 'unattainability'. Anything 'unattainable' can never be reached.
13. In Problem 49 of the Ahmose Papyrus of ancient Egypt (c. 1650 BC) appears a labelled diagram showing a square with four isosceles triangles removed leaving an octagon which is interpreted to have an area equal to that of a circle inscribed in the square. For further details, see Joseph (2000: 82–83).
14. In terms of modern notation, this relation may be expressed as

$$x_0 = 1,\ x_{n+1} = \frac{\sqrt{1+x_n^2}-1}{x_n},\quad \pi = 4\lim_{n\to\infty} 2^n x_n$$

15. In terms of modern notation

$$\pi = 4\lim \frac{1}{4}\sum_{n=1}^{N}\left[\frac{1}{1+\left(\frac{n}{N}\right)^2}\right] \text{ where the sum tends towards } \int_0^1 \frac{dx}{1+x^2}$$

Taking the limit of the above series gives the infinite series.

$$C = 4D\ 1-\frac{1}{3}+\frac{1}{5}-\frac{1}{7}+\ldots$$

for the circumference of a circle C of diameter D.

16. According to the standard histories of mathematics, the Pythagorean Greeks knew (I) in the sixth century BC. In the third century BC, Archimedes in his book *On Spirals* gave a 'formula' for the sum of squares whose equivalent modern formulation is given in (II). Archimedes applied this to deduce the area inside what we now call an Archimedean spiral by the classical Greek method of exhaustion. The ability to sum yet higher powers was key to finding areas and volumes of other geometric objects. We find a number of works in which there was an understanding of the method of finding sums of cubes, including the works of Nicomachus of Gerasa (first century BC), Aryabhata in India (AD 499) and al-Karaji in the Arab world (c. AD 900–1000). The first evidence of a general relationship between various exponents is in the work of ibn al-Haytham (965–1039), who needed a formula for a sum of fourth powers in

order to find the volume of a general paraboloid of revolution. Although not stated in full generality, his discovery was essentially the recursive relationship

$$(n+1)\sum_{i=1}^{n} i^k = \sum_{i=1}^{n} i^{k+1} + \sum_{p=1}^{n}\left(\sum_{i=1}^{p} i^k\right)$$

For further details, see Berggren (2007: 587–92).

17. This is a point rarely noted or commented on in the discussion of Kerala mathematics. I am grateful to Prof. Narasimha for drawing my attention to this fact.
18. The basic approach is to inscribe or circumscribe a regular polygon. The problem then is to find the side of the polygon as a multiple of the diameter. An interesting extension of the method, first suggested by Eudoxus of Cnidus (fl. 375 BC) and then extensively used by Archimedes, was a rigorous alternative to 'taking the limit' which the Greeks avoided given their well-known 'horror of the infinite'. It is based on the simple observation that if a circle is enclosed between two polygons of *n* sides, then, as *n* increases, the gap between the circumference of the circle and the perimeters of the inscribed and circumscribed polygons diminishes so that eventually the perimeters of the polygons and the circle would become identical. Or, in other words, as *n* increases, the difference in the area between the polygons and the circle would be gradually exhausted.
19. In the sixth chapter of Jyesthadeva's *Yuktibhasa*, there is a section entitled 'Circumference of a Circle without Calculating Square-roots' that contains a detailed description of this approach (Sarma 2008: 49–68).
20. The special importance of the Pythagorean result in Indian mathematics is seen from the presentation of a proof at the beginning of the crucial sixth chapter of the *Yuktibhasa* on the 'Circumference' that leads to the infinite series for π. The proof presented is what we would now describe as 'dissect and re-assemble'. For further details, see Sarma (2008: 44–45, 179–80).
21. Verse 26 was given in an earlier chapter and may be translated as, 'In the Rule of Three, multiply the *pramana-phala* (fruit or ***p***) by the *iccha* (desire or ***d***) and divide by the *pramana* (measure/argument or ***a***). The result is the fruit of desire (***f***), i.e., $f = \frac{pd}{a}$, (Shukla 1976)

    A whole short chapter (Chapter 4) in the *Yuktibhasa* was devoted to a detailed consideration of 'Rule of Three', including the 'reverse Rule of Three'. The *phala* in 'reverse of three' is got by dividing, by the *iccha*, the product of *pramana* and *pramana-phala*, that is, $f^* = \frac{ap}{d}$. For further details, see Sarma (2008: 28–30).

    The chapter concludes: 'Most of mathematical computations are pervaded by the "Rule of Three" (*trairasika-nyaya*) and the "rule of base, height and hypoteneuse" or the so-called Pythagorean theorem (*bhuja-koti-karna-nyaya*).' For further details, see Sarma (2008: 28–30).
22. A familiar result in elementary algebra relates to the infinite decreasing geometric series $1 - x + x^2 - x^3 + \ldots$ where $x < 1$ is the common ratio with the sum equal to

$$\frac{1}{1-(-x)} = \frac{1}{1+x}$$

23. This asymptotic relation made its first appearance in Europe in the works of Roberval in 1634 and Fermat in 1636.
24. Note that this is equivalent to $\frac{\pi}{4}=1-\frac{1}{3}+\frac{1}{5}-\frac{1}{7}+\ldots$
25. It is interesting in this context to note about 500 years after Madhava, Yesudas Ramchandra wrote a book in 1850, entitled *A Treatise on the Problems of Maxima and Minima* in which he claimed that he had developed a new method, consistent with the Indian tradition of mathematics, to solve all problems of maxima and minima by algebra and not calculus. This book was republished in England with the help of the British mathematician, Augustus De Morgan. For further details, see Joseph (1995).
26. The derivation that follows is based on the seminal text of Kerala mathematics, the *Yuktibhasa*, composed by Jyesthadeva around 1530. It is not a commentary on the *Tantrasangraha* but rather 'its aim is to provide the basic equipment needed by one who desires to study the computation of planetary movements as depicted in the *Tantrasangraha*'. The purpose it serves is to introduce basic concepts and theories of mathematics and astronomy, providing the definitions, and setting out the methodologies and their rationales. It is unique in two respects. Unlike most primary texts in Indian mathematics, *Yuktibhasa* was written in the local language Malayalam; and as an expository text it contains detailed rationale and proofs of various results. Appendix I of this chapter provides an introduction to this text.
27. As we saw earlier, with 19 terms on the right-hand side, the ratio of circumference over diameter (π) does not achieve even a one decimal place accuracy.
28. The relevant portions from the *Yuktibhasa* where these subjects are discussed are in Sarma (2008, Volume 1).
29. A variant of this method was already discussed in the Appendix in Chapter 4.
30. In an attempt to discover an infinite series representation of any given trigonometric function and the relationship between the function and its successive derivatives, Gregory stumbled on the arctan series. He took, in terms of modern notation, $d\theta=\frac{d(\tan\theta)}{1+\tan^2\theta}$ and carried out term by term integration to obtain his result: a procedure not dissimilar to the modern derivation given earlier. Leibniz's discovery arose from his application of fresh thinking to an old problem, namely quadrature of a circle. In applying a transformation formula (similar to the present-day rule for integration by parts) to the quadrature of the circle, he discovered the series for π. It must be pointed out, however, that the ideas of calculus such as integration by parts, change of variables and higher derivatives were not completely understood then. They were often dressed up in geometric language with, for example, Leibniz talking about 'characteristic triangles' and 'transmutation'.
31. In Indian mathematics, there has been a singular absence of 'deductive proof' based on a set of 'self-evident' axioms in the style of Euclid. Although such 'formal proofs' are not part of the Indian approach, demonstrations, explanations and search for rationales have been preoccupations for a long time. As such, geometric intuition and computational reasoning formed the basis of 'proofs' both in India and later in Europe. By the end of the nineteenth century, geometry fell out of favour to be

replaced by arithmetic and set theory. Thus Bolzano and Dedekind tried to prove that infinite sets exist by arguing that any object of thought can be thought about and thus give rise to a new thought object. Today we reject such proofs and use an axiom of infinity. Starting from a practical orientation and serving practitioners of the astronomical arts, the subject of analysis by its peculiar logic developed eventually into a highly abstract and rarefied entity for the delectations of primarily the professional mathematician.

32. It must be emphasised that the *Yuktibhasa* is not a commentary on *Tantrasangraha* like Sankara Variyar's *Yuktidipika*. At the beginning of one of the versions of the *Yuktibhasa* there is the statement: 'Here, commencing an elucidation in full of the rationale of planetary computations according to the *Tantrasanghraha* …' (Sarma 2008: xxxii).
33. *Kali* day (*ahargana*) is the number of civil days that have elapsed since the beginning of *Kaliyuga*. The calculation is shown in Sarma (2008: 31–32, 170–71).
34. In the *Yuktibhasa*, $F_a(n)$ appears only as a step in the derivation of $F_b(n)$. Only $F_b(n)$ and $F_c(n)$ are given explicitly. The expressions of $F_b(n)$ and $F_c(n)$ as continued fractions reveal the fact that the incorporation of $F(n)$ permits an interpretation of the three correction terms as the first three convergents of an infinite continued-fraction. We will not examine this interpretation here. For a brief discussion of this point, see Rajagopal and Rangachari (1986: 90) and Sarma (2008: 82, 207).
35. See Hayashi et al. (1990) for further details.
36. For more details, see Sarma (2008: 80–82, 205–6).
37. (9) and (10) are based on the following passages from the *Yuktibhasa:*

> Multiply the square of the diameter by 12 and extract the square root of the product. This is the first term of a sequence in which each of the successive terms is divided by the odd numbers 1,3,5,… and then from the second term onwards a further division by successive powers of 3 produces the final terms of this sequence. The series is now formed by subtracting and adding alternately the successive terms, starting from the second term. (Sarma 2008: 70)

> Divide four times the given diameter by the cubes of the odd numbers 3,5,7,9,… subtracting from each its respective roots. The terms thus obtained alternately added to and subtracted from three times the diameter. This is the circumference of the circle whose diameter is given. (Sarma 2008: 81)

38. Consider other examples:
    a. In the *Tantrasangraha*, Nilakantha writes: 'Multiply any given diameter by 104348 and divide the product by 33215, the quotient is a very correct circumference.' This is the same as taking

       $\pi = 3.131592653921$ which is correct to 10 decimal places.

    b. In *Kriyakramakari*, Madhava is reported to have said: 'for a diameter of 900,0000,0000 (9×10^{11}) the circumference is 28,27,43,33,88,233'. The value of π implied is correct to 11 decimal places.

c. In *Karanapaddhati* (AD 1732), it is stated: 'If the circumference of a circle in minutes is multiplied by 10,000,000,000 (10^{11}) and the product divided by 3,141,59,26,536, the result will be the diameter of the circle in terms of the minutes of the circumference, and its half will be the radius.' The implied value of π is correct to 10 decimal places. (For details see Footnote 3 on Page 26 of Sarma 1972).

7

Reaching for the Stars: The Power Series for Sines and Cosines

Introduction

In a commentary on Nilakantha's *Tantrasangraha*, by an unknown student of Jyesthadeva, the author of *Yuktibhasa*, are found the following descriptions of the power series for sine and versine without any derivations.[1]

(A) 'The arc is repeatedly multiplied by the square of itself and divided (in order) by the square of each and every even number increased by itself and multiplied by the square of the radius. The arc and the terms obtained from these repeated operations are to be placed one beneath the other in order, and the last term subtracted from the one above, the remainder from the term then next above, and so on, to yield the *[bhuja] jya* (or Indian Sine) of the arc.'

(B) 'The radius is repeatedly multiplied by the square of the arc and divided (in order) by the square of each and every even number diminished by itself and multiplied by the square of the radius, with the first term involving only 2. The resulting terms are placed one beneath the other in order, and the last term subtracted from the one above, the remainder from the term next above and so on, to yield *ukramajya* or *sara* (Indian versine) of the arc.' (Rajagopal and Rangachari 1978)

Expressed symbolically, where r is the radius and a is the length of the given arc, the first three even numbers given by (A) is equivalent to:

$$\frac{a\,.\,a^2}{1!(2^2+2)r^2}=\frac{a^3}{3!r^2}$$

$$\frac{a^3.\,a^2}{3!(4^2+4)r^4}=\frac{a^5}{5!r^4}$$

$$\frac{a^5.\,a^2}{5!(6^2+6)r^6}=\frac{a^5}{7!r^6}$$

Hence, Indian Sine $= r\sin\theta = a-\dfrac{a^3}{3!r^2}+\dfrac{a^5}{5!r^4}-\dfrac{a^7}{7!r^6}+\ldots$

Substituting $a/r = \theta$ gives:

$$\sin\theta=\theta-\frac{\theta^3}{3!}+\frac{\theta^5}{5!}-\frac{\theta^7}{7!}+\ldots \qquad (1)$$

Using the foregoing notation and denoting Indian Versine (*sara*) by $(r - r\cos\theta)$, the first three even numbers given in (B) can be written as:

$$\text{Indian Versine } (r - r\cos\theta) = \frac{a^2}{2!r}-\frac{a^4}{4!r^3}+\frac{a^6}{6!r^6}$$

Substituting $\dfrac{a}{r}=\theta$ and simplifying gives:

$$\cos\theta=1-\frac{\theta^2}{2!}+\frac{\theta^4}{4}-\frac{\theta^6}{6!}+\ldots \qquad (2)$$

The series given in (1) and (2) are usually named after Newton. They make their first appearance in European mathematics in a letter from Newton to Oldenburg in 1676 and then provided on a firmer algebraic basis by De Moivre (1708–1738) and Euler (1748). The series should be more appropriately named after Madhava to whom the series are usually attributed to by the later members of the Kerala School.

The Background and Alternative Formulations

The seventh chapter of *Yuktibhasa* (Sarma 2008: 84–86) contains a discussion of the sine and cosine of a circle. Starting with two well-known results from the Aryabhatan era, namely:

(i) $r\sin \pi/6 = \frac{1}{2}r$ and (ii) $r\sin\pi/2 = r$

and applying the formulae

$$r\sin\theta = \frac{1}{2}\sqrt{(r\sin 2\theta)^2 + r^2(1-\cos 2\theta)}^{2} \quad \text{and} \quad r\sin\left(\frac{\pi}{2}-\theta\right) = \sqrt{r^2-(r\sin\theta)^2}$$

the values of $r\sin\frac{c\pi}{48}$ are generated for c = 1, 2, …, 24. Using yet another known result: the cosine of an arc of a circle is equal to its complementary arc (that is, the portion of the arc required with the given arc to complete a quadrant), the values of $r\cos\frac{c\pi}{48}$ are also generated for c = 1, 2, …, 24. A problem arises in calculating the sine and cosine of an arc whose length is not an exact multiple of $\frac{\pi r}{48}$. The length of an arc can lie anywhere between 0 and $2\pi r$. It was known then that in order to obtain the sine and cosine of an arc of any length, it was sufficient to evaluate the sine and cosine of the portion of the arc that lies in a quadrant of the circle.[2] It was the search for finding the length of any arc (which were not exact multiples of $\frac{\pi r}{48}$) lying in the first quadrant that led to the work on infinite sine and cosine series. In the next section of this chapter we will examine how these series were derived in Kerala.

It should be noted that the main motivation behind the evaluation of arc-lengths was primarily astronomical, as can be seen from the following representation of the sine and versed-sine series apart from the ones given in the last section. In Nilakantha's *Aryabhatiyabhasya* (and repeated in the *Yuktibhasa*) appears the following passage, again attributed to Madhava:

> For a sequence of five numbers (3, 5, 7, 9, 11), the first number is multiplied by the square of the given arc and divided by the square of 5400. The quotient is subtracted from the second number. The result of the subtraction is next multiplied by the square of the given arc and divided by the square of 5400 and the quotient that results is subtracted from the third number. This process of multiplication and division followed by subtraction is repeated till all the five numbers are gone through. The final result is then multiplied by the cube of the given arc and divided by the cube of 5400. If the quotient then resulting is subtracted from the given arc, what remains will be the required *[bhuja] jya* (or Indian sine). (Rajagopal and Rangachari 1978: 97–98)

Expressed in symbolic notation and with the following additional symbols, where all angular measures are given in radians:

Let $c = 90 \times 60 = 5400$ *ilis* ('minutes') be the first quadrant of a circle of circumference 21600 *ilis*; a_i be the ith odd number where $i = 3, 5, 7, 9, 11$; and $s = r\theta$ be the sine-chord.

Then

$$jya\theta = r\sin\theta = s - \left(\frac{s}{c}\right)^3 \left\langle a_3 - \left(\frac{s}{c}\right)^2 \left\{ a_5 - \left(\frac{s}{c}\right)^2 \left[a_7 - \left(\frac{s}{c}\right)^2 \left(a_9 - \left(\frac{s}{c}\right)^2 a_{11} \right)\right]\right\}\right\rangle$$

Or

$$jya\theta = r\sin\theta = s - a_3\left(\frac{s}{c}\right)^3 + a_5\left(\frac{s}{c}\right)^5 - a_7\left(\frac{s}{c}\right)^7 + a_9\left(\frac{s}{c}\right)^9 - a_{11}\left(\frac{s}{c}\right)^{11}$$

Now substitute the individual values of $a_i = 3, 5, 7, 9, 11$; $c = 5400$ and $s = r\theta$ in the above expression and simplify to get the power series for sineθ

$$\sin\theta = \theta - \frac{\theta^3}{3!} + \frac{\theta^5}{5!} - \frac{\theta^7}{7!} + \frac{\theta^9}{9!} - \frac{\theta^{11}}{11!}\ldots$$

which is the same as (1) in the last section.

The verse that follows in the *Yuktibhasa* gives the cosine series:

> For a sequence of the first six even numbers (i.e, 2, 4, 6, 8, 10, 12), the first number is multiplied by the square of the given arc and divided by the square of 5400. The quotient is subtracted from the second number. The result of the subtraction is next multiplied by the square of the given arc and divided by the square of 5400 and the quotient that results is subtracted from the third number. This process of multiplication and division followed by subtraction is repeated till all the six numbers are gone through. The final result will be the required *utkramajya* or *saram* (Indian versine) of the arc. (Rajagopal and Rangachari 1978: 98–99)

Or, denoting the first six even numbers as a_j where $j = 2, 4, 6, 8, 10, 12$

$$r - r\cos\theta = \left(\frac{s}{c}\right)^2 \left\langle a_2 - \left(\frac{s}{c}\right)^2 \left\{ a_4 - \left(\frac{s}{c}\right)^2 \left[a_6 - \left(\frac{s}{c}\right)^2 \left(a_8 - \left(\frac{s}{c}\right)^2 a_{10} \right)\right] \ldots \right\}\right\rangle$$

$$r - r\cos\theta = (s/c)^2[a_2 - (s/c)^2\{a_4 - (s/c)^2(a_6 - (s/c)^2[a_8 - (s/c)^2[a_{10} - (s/c)^2 a_{12}\}]$$

Or more generally

$$r - r\cos\theta = a_2(s/c)^2 - a_4(s/c)^4 + a_6(s/c)^6 - a_8(s/c)^8 + a_{10}(s/c)^{10} - a_{12}(s/c)^{12} + \ldots$$

Substitute the values of a_j's and $s = r\theta$ in the above and simplify to get:

$$\cos\theta = 1 - \frac{\theta^2}{2!} + \frac{\theta^4}{4!} - \frac{\theta^6}{6!} + \frac{\theta^8}{8!} - \frac{\theta^{10}}{10!} + \frac{\theta^{12}}{12!} - \ldots$$

which is the same as (2) in the last section.

In Indian geometry of the time, the circumference of a circle was divided into $360 \times 60 = 21600$ *ilis* (or 'minutes'). An *ili* of an arc faces an angle of one minute at the centre of a unit circle.[3] In terms of *ilis*, the radius is reckoned as $3437'44''43'''$.[4]

Now $\quad C/4 = 5{,}400 = \tfrac{1}{2}\pi r$

Define[5] $\quad a_1 = 5{,}400 = r(\tfrac{1}{2}\pi)$

$$a_2 = \frac{(5,400)^2}{2!r} = \frac{r}{2!}\cdot(\tfrac{1}{2}\pi)^2$$

$$a_3 = \frac{(5,400)^3}{3!r} = \frac{r}{3!}\cdot(\tfrac{1}{2}\pi)^3$$

$$a_4 = \frac{(5,400)^4}{4!r} = \frac{r}{4!}\cdot(\tfrac{1}{2}\pi)^4$$

$$a_5 = \frac{(5,400)^5}{5!r} = \frac{r}{5!}\cdot(\tfrac{1}{2}\pi)^5$$

$$a_6 = \frac{(5,400)^6}{6!r} = \frac{r}{6!}\cdot(\tfrac{1}{2}\pi)^6$$

$$\vdots$$

$$a_{12} = \frac{(5,400)^{12}}{12!r} = \frac{r}{12!}\cdot(\tfrac{1}{2}\pi)^{12}$$

It is believed that Madhava used this result to construct a sine table of 24 values for θ (i.e., the values of cumulative series obtained by dividing the quadrant of a circle into 24 equal parts) starting from $\theta = 224$ *ilis*, 50 *vilis* and 22 *tatparas* which corresponds to the present-day angular measurement of 3 degrees and 45 minutes. The values of his sine table are correct in almost all cases to the 8th or 9th decimal place. A similar development may have led to the construction of an equally accurate cosine table.

Derivations of Sine and Cosine Series

In Figure 7.1, the arc PX of a circle of radius r subtends an angle θ at the centre O. Let the arc PX be of length x and let $PP_1 = \delta x$ be a small increase in the arc length with corresponding increase of angle $\delta\theta$. P_m is the mid-point of the arc PP_1. Q_1, Q_m and Q are perpendiculars dropped from P_1, P_m and P respectively to the radius OX.

Applying the 'Rule of Three' (or properties of similar triangles) to triangles P_1SP and OQ_mP_m we have:

$$\delta(\mathrm{r}\cos\theta) \approx \frac{-\mathrm{r}\sin\theta\,\delta\mathrm{x}}{\mathrm{r}} \qquad (3)$$

In what follows we continue to translate the exposition in the *Yuktibhasa* into modern mathematical language. So if $r = 1$, then $\delta x \approx \delta\theta$ and the above relationship expressed in (3) would reduce to

$$\delta(\sin\theta) \approx \cos\theta\,\delta\theta \qquad (3a)$$

$$\delta(\cos\theta) \approx -\sin\theta\,\delta\theta \qquad (3b)$$

Figure 7.1

The Derivation of Sine and Cosine

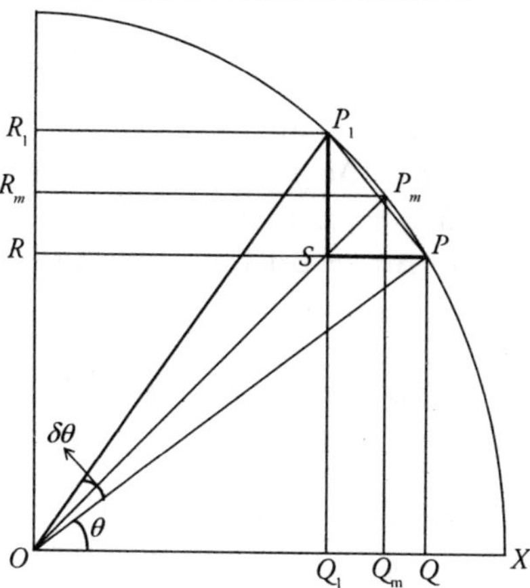

The formulae $\sin\theta$ and $\cos\theta$ in the *Yuktibhasa* utilises the foregoing derivatives. However, we have already seen in the mathematical exposition so far that there is a marked difference between Indian and modern mathematics. This not only applies to epistemology but also to philosophy and to terminology. In order to present an exposition that a reader versed only in modern mathematics can make sense we have 'translated' the mathematics of the *Yuktibhasa* using modern symbolic language. However, this will not invalidate the argument that the geometric and analytical thinking underlying Kerala mathematics has no parallel in Western mathematics—their treatment of infinity and infinitesimal processes is special case in point. All this is even more marked in their introduction and use of asymptotic expansions. All this is even more marked in construction of infinite series expansions. Consequently, a full appreciation of this analysis in the *Yuktibhasa* can only be achieved by bearing these differences in mind. This point is emphasised in the works of both Saraswati Amma (1963) and Bag (1976) and by Srinivas in Sarma (2008).

In Figure 7.2, the radius of the circle is r. Let the arc PX of length x, which subtends an angle θ at the centre, be subdivided into x equal parts each of length δx = one unit of measure. Let σ_i and $\Delta\sigma_i$, $1 \le i \le x$, be the sine-chords and sine-differences, respectively, at these points. Let χ_i,

$1 \le i \le x$, be the cosine-chords at the mid-points of the arcs.

Now $\Delta\sigma_1 = \sigma_1$

So the sum of the difference of the sine-differences

$$\begin{aligned}
&= [(\Delta\sigma_1 - \Delta\sigma_2) + (\Delta\sigma_1 - \Delta\sigma_3) + \ldots + (\Delta\sigma_1 - \Delta\sigma_x)] \\
&= [(\sigma_1 - \Delta\sigma_2) + (\sigma_1 - \Delta\sigma_3) + \ldots + (\sigma_1 - \Delta\sigma_x)] \\
&= [n\sigma_1 - (\Delta\sigma_2 + \Delta\sigma_3 + \ldots + \Delta\sigma_x)] \\
&= [(\Delta\sigma_1 - \Delta\sigma_2) + (\Delta\sigma_1 - \Delta\sigma_3) + \ldots + (\Delta\sigma_1 - \Delta\sigma_x)] \\
&\approx r\theta - \sigma_x = r\theta - r\sin\theta
\end{aligned} \tag{4}$$

Now the height H (= PM) of the arc PX = sum of the positive decreases in cosines. So using (3) we have

$$H = \frac{(\sigma_1\delta x + \sigma_2\delta x + \sigma_3\delta x + \ldots + \sigma_x\delta x)}{r} \tag{5}$$

Figure 7.2

The Derivation of Sine and Cosine Series

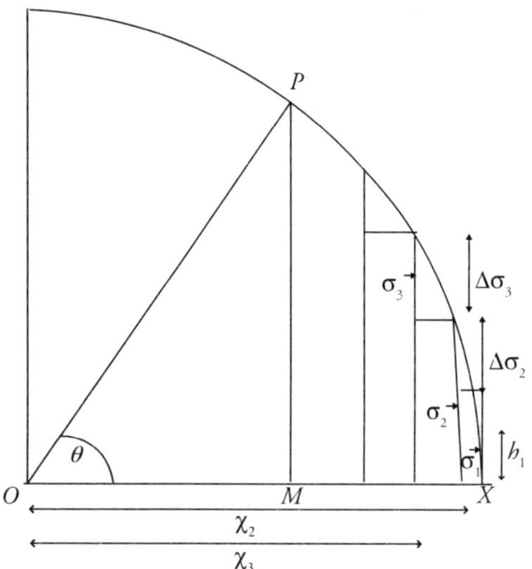

A first approximation is to take $\sigma i = i\delta x = i$ (as $\delta x = 1$), $1 \le i \le x$. That is, *the sine-chords are assumed to be* ***equal*** *to the lengths of the arcs.* Noting that this is an over-estimation, a first approximation for H in (5) is tried:

$$H = \frac{(1+2+3+...+x)}{r} \tag{5a}$$

Now consider the following approximation

$$1^k + 2^k + \ldots + x^k \approx x^{k+1}/(k+1) \tag{6}$$

Using (6) with $k = 1$ in (5a) gives

$$H = \frac{(1+2+3+...+x)}{r} \approx x^2/2r \tag{7}$$

On the other hand, the sine-differences can be calculated another way. Using (3a) we can write

$$\Delta\sigma_1 \approx \chi_1\ \delta x = \chi_1 \text{ (as } \delta x = 1)$$

Now the *Yuktibhasa* makes the approximation $\chi_1 \approx (r - h_1)$, where h_1 represents the height at the mid-point of the first arc element (shown in Figure 7.2). So that

$$\Delta\sigma_1 \approx (r - h_1) \tag{8a}$$

Similarly,

$$\Delta\sigma_i \approx \chi_i \delta x \approx (r - h_i),\ 1 \leq i \leq x, \tag{8b}$$

where h_i represents the height at the mid-point of the ith arc element.

Thus $\quad (\Delta\sigma_1 - \Delta\sigma_i) \approx (h_i - h_1),\ 1 \leq i \leq x, \quad$ (8c)

The *Yuktibhasa* states that it is reasonable to assume that h_1 is approximately equal to 0 and h_x is approximately equal to H (the height of the arc PX).

Thus $\quad (\Delta\sigma_1 - \Delta\sigma_x) = h_x = H \approx x^2/2r \quad$ (9)

Similarly,

$$(\Delta\sigma_1 - \Delta\sigma_i) = h_i = \approx (x - i)^2/2r,\ 1 \leq i \leq x, \tag{10}$$

So that $\quad \sum_1^x (\Delta\sigma_1 - \Delta\sigma_i) = \sum_1^x (x-i)^2/2r \quad$ (11)

From (4) and (7) with $k = 2$ we have

$$r\theta - r\sin\theta \approx {}^1/_2 \times (x^3/3r^2) = x^3/3!r^2 \tag{12}$$

Initially, the sine-chords were assumed to be *equal* to the lengths of the arcs (when, in fact, they are less than the lengths of the arcs). The excess of the arc length over the sine-chord implied by the foregoing is now applied to the right side of equation (8):

$$H = \frac{(1+2+3+\ldots+x)}{r} = \frac{(x+(x-1)+(x-2)+\ldots+1)}{r}$$

so that the second approximation is

$$H = \frac{1}{r} \times [(x - x^3/3!r^2) + (x-1-(x-1)^3/3!r^2) + (x-2-(x-2)^3/3!r^2) + \ldots)]$$

$$= \frac{(x+(x-1)+(x-2)+\ldots)}{r} - \frac{[x^3/3!r^2+(x-1)^3/3!r^2+(x-2)^3/3!r^2+\ldots)}{r}$$

Thus the correction for the height $H =$

$$= \frac{[x^3/3!r^2+(x-1)^3/3!r^2+(x-2)^3/3!r^2+\ldots]}{r}$$

Or $\quad rH = x^4/4!r^3$; using (7) with $k = 3$ (13)

With this correction we now have

$$\sum_1^x(\Delta\sigma_1 - \Delta\sigma_i) = \sum_1^x (x-i)^2/2r - \sum_1^x (x-i)^4/4!r^3$$

Using (6) with $k = 2$ and $k = 4$ and (12) we have

$$r\theta - r\sin\theta \approx x^3/3! - x^5/5! \tag{14}$$

In the next step, the correction term is introduced to correct the deficiency of the approximation for the sine-chord. So that the next approximation is

$$r\theta - r\sin\theta \approx x^3/3! - x^5/5! - x^7/7! \tag{14a}$$

Continuing in this way, ad infinitum, we have

$$r\theta - r\sin\theta = x^3/3! - x^5/5! + x^7/7! - \ldots + (-1)^n x^{2n-1}/(2n-1)! + \ldots \tag{15}$$

Substituting $x = r\theta$ gives the familiar series

$$\sin\theta = \theta - \theta^3/3! + \theta^5/5! - \theta^7/7! + \ldots + (-1)^n\theta^{2n+1}/(2n+1)! + \ldots \tag{16}$$

The familiar infinite series for $\cos\theta$ is constructed in likewise manner. It should be remembered that these developments were first expounded in the fifteenth century by Madhava. However, in the vast majority of published books on the history of mathematics, the series expansions for $\sin\theta$ and $\cos\theta$ are attributed to Newton in his works in 1669 (*Analysis with Infinite Series*) and in 1671 (*Method of Fluxions and Infinite Series*).

Rational Approximations of Sine and Cosine Functions

In Nilakantha's *Aryabhatiyabhasya* (page 73) and Jyesthadeva's *Yuktibhasa* (page 178) appears the following passage with '*iti Madhava*' ('thus said Madhava') appearing at the end of the verse in the latter text: Expressed symbolically, let h be the arc-difference. The 'divisor' (d) = 13,751/2h.

Then

$$\sin x + (\cos x - \sin x/d)(2/d) \approx \sin(x + h)$$

And

$$\cos x - (\sin x + \cos x/d)(2/d) \approx \cos(x + h)$$

Now

$$d = 2r/h \text{ since } r = \frac{21{,}600}{2\delta} = 3437.75\delta = \frac{13{,}751}{4}$$

Substitute $d = 2r/h$ in the expressions to get

$$\sin(x + \delta h)\ \ \sin x + (h/r)\cos x - (h^2/2r^2)\sin x$$
$$\cos(x + \delta h)\ \ \cos x - (h/r)\sin x - (h^2/2r^2)\cos x$$

These results are but special cases of familiar expansions in mathematics, the Taylor series, up to the second power of the small quantity u

$f(x + u) = f(x) + u.f'(x) + (u^2/2)f''(x) + \ldots$ where u (in radian measure) $= h/r$

Construction of the Sine Tables: The Kerala Contribution

As discussed in a previous chapter, over many years, Indian mathematicians have made notable contributions to trigonometry under the caption '*jyotpatti*' (*jya* + *utpatti* = source of RSines). This branch of mathematics evolved from astronomical needs, such as computation of latitudes or of position of planets, their movements, and so on. As Nilakantha pointed out in his *Golasara*, while explaining the concept of *jyā* (or RSines), that he was computing sines and cosines because they were required for a discussion of the motion of planets in their respective orbits on the stellar sphere.[6]

As a case study of the advances made in trigonometry by the Kerala School, let us re-consider Nilakantha method of computing sine tables as shown in his *Golasara*.[7] This has already been discussed in an earlier chapter. Here, we are concerned with a summary statement of the methodology used by Nilakantha.

The procedure for computation of sine and cosine tables of length $l = 3^{*}\ 2^{m}$ at equal arc intervals $h = \frac{1800}{l}$ for $m = 0, 1, 2, 3, 4, 5$, and so on, consists of the following steps.

Step I: Take $R = \frac{21600 \times 113}{2 \times 355}$ computed using $\pi = \frac{355}{113}$ as stated in the rule.

Step II: Start with, $\theta = 30°$, $S_0 = RSin30° = \frac{R}{2}$

and $C_0 = R\text{Cos}30° = \sqrt{R^2 - S_0^2}$

Step III: Denote $S_i = R\text{Sin}\left(\frac{\theta}{2^i}\right)$ and $C_i = R\text{Cos}\left(\frac{\theta}{2^i}\right)$ and calculate

$$S_i = \sqrt{S_{i-1}{}^2 + (R - C_{i-1})^2} \text{ and } C_i = \sqrt{R^2 - S_i^2}, \text{ for } i = 1, 2, 3, \ldots, m$$

where $m = 3, 4, 5, \ldots$ according as the arc interval h is 225′, 112.5′, 56.25′, …

Step IV: Now for initiating the computation of RSine and RCosine tables at interval of $h = \frac{\theta}{2^m} = \frac{30 \times 60}{2^m}$, the first tabular RSine $= J_1$ is the value S_m and the last RSine is $J_l = R\text{Sin}lh = R$, the half diameter, where

$$l = \frac{90 \times 60}{\left(\frac{\theta}{2^m}\right)} = \frac{90 \times 60 \times 2^m}{30 \times 60} = 3 \times 2^m, \text{ since } \theta = 30° = 30 \times 60'$$

Step V: Compute $J_{l-1} = \sqrt{R^2 - J_1^{\,2}}$, $\Delta J_{l-1} = J_l - J_{l-1}$ and

$$\lambda = 2\left(\frac{\Delta J_{l-1}}{R}\right)$$

Step VI: Compute $\Delta J_{l-k} = \lambda \times J_{l-(k-1)} + \Delta J_{l-(k-1)}$

Step VII: Now for $k = 2, 3, 4, \ldots l-2$ compute $J_{l-k} = J_{l-(k-1)} - \Delta J_{l-k}$ and

$$C_{l-k} = \sqrt{R^2 - J_{l-k}^{\,2}}$$

Sine tables of lengths 3, 6, 12, 24, 48, 96, 192, 384, ... may be computed using this *Golasara* algorithm. It is quite interesting to note that Nilakantha has referred to the last and the first RSine-differences by the terms *antya* and *adi khanda* without mentioning that the last RSine is the 24th. So it may be inferred that Nilakantha's rule for determination of the RSines successively gives a general method for constructing sine and cosine tables. By comparing the values obtained with the corresponding modern values it can be shown that the *Golasara* method gives by and large fairly accurate values even up to 20 decimal places computed. Cosine tables may also be constructed similarly.

Conclusion

A historical and comparative study of infinite service provides a suitable vehicle for testing certain perceptions about different mathematical traditions. A widely accepted view among historians of mathematics is that mathematics outside the sphere of Greek influence, such as Indian or Chinese mathematical traditions, was algebraic in inclination and empirical in practice which provided a marked contrast to Greek mathematics which was geometric and anti-empirical. Again, many of the commonly available books on history of mathematics declare or imply that Indian mathematics, whatever be its other achievements, did not have any notion of proof. What a comparative study would indicate are the dangers of such categorisation and generalisation. And in a deeper sense it would bring home the point that between different mathematical traditions there are certain basic differences in the cognitive structures of mathematics, such as differences in their ontological conceptions regarding the existence and nature of mathematical objects and differences in methodological conceptions regarding the nature and ways of establishing mathematical truths.

Notes

1. No information is available in the text on the ultimate source of these results, although there is extraneous evidence to indicate that they originated with the founder of the Kerala School, Madhava.
2. What was known may be summarised in modern notation as
$$\sin(90^\circ + \theta) = \sin(270^\circ + \theta) = -\cos\theta$$
$$\sin(180^\circ + \theta) = -\sin\theta$$
The cosines of quantities outside the first quadrant may be evaluated similarly.
3. An *ili* is further subdivided into 60 *vili*s (or *vikalas*) and a *vili* is further subdivided into 60 *talpara*s and a *talpara* into 60 *pratalaparas*.
4. Or more precisely, the radius is taken as 3437 *ili*s, 44 *vili*s, 48 *talpara*s and 22 *pratalpara*s (or in decimal notation 3437.74677) which is approximately equal to the radian measure, 57 degrees, 17 minutes and 44.8 seconds correct to a *talpara* (or 3437.77 in decimal notation).
5. In the original text of *Yuktibhasa* from which the passage is quoted later, each a_i ($i = 1, 2, \ldots, 12$) is assigned an individual name according to the *Katapyadi* notation and the associated numerical values. Thus, for example, the last three coefficients, a_{10}, a_{11} and a_{12}, were given the names *stripsuna* (52 *ili*s), *vidvan* (44 *vili*s) and *stena* (6 *vili*s) respectively.
6. See *Golasara Siddhantadarpaṇamca* of *Gargya Kerala Nilakaṇṭha*, Ms No. T 846. B, Transcript copy by Paramesvara Sastry, C.1024.E (K.U.O.R.I and Mss Library, Trivandrum), iii vs.2
7. For details see Mallayya (2004).

8

Changing Perspectives on Indian Mathematics*

Introduction: Judgements on Indian Mathematics

In the early years of the second millennium AD, the more significant evaluations of Indian mathematics and astronomy were those from the Islamic world. Some, like Said al-Andalusi (1068), claimed them to be of exceptionally high order:

> [The Indians] have acquired immense information and reached the zenith in their knowledge of the movements of the stars [astronomy] and the secrets of the skies [astrology] as well as other mathematical studies. After all that, they have surpassed all the other peoples in their knowledge of medical science and the strengths of various drugs, the characteristics of compounds, and the peculiarities of substances. (Andalusi [c.1068] translated by Salem and Kumar 1991: 11–12)

Others like al-Biruni were more measured. His overall assessment of Indian mathematics and astronomy is similar to a contemporary

*This chapter is based on two recent papers by Almeida and Joseph (2004, 2009b) which were the outcomes of their collaboration on a research project funded by Arts and Humanities Research Board, United Kingdom. The author owes a considerable debt to Dennis Almeida for his contributions which were crucial to the success of the project.

pessimistic assessment of the vast mathematical literature of the twenty-first century—uneven with a few good quality research papers and a majority of nondescript or even error-strewn publications. Al-Biruni (1030): translated by Ahmad 1999: 11–12) wrote: 'I can only compare their [i.e., Indian] mathematical and astronomical literature, as far as I know it, to a mixture of pear shells and sour dates, or of pearls and dung, or of costly crystals and common pebbles.'

Nevertheless, a common element in these early evaluations is the insistence on the uniqueness of Indian mathematics. By the nineteenth century, however, and contemporaneous with the rise of European colonisation in the East, the views of European scholars regarding the supposed superiority of European knowledge had developed ethnocentric overtones. In 1873, Sedillot asserted that not only was Indian science indebted to Europe but also that the Indian numbers were an 'abbreviated form' of Roman numbers and that Sanskrit was merely 'muddled' Greek. Although Sedillot's assertions may be judged today as being based on imperfect knowledge and understanding of the nature and scope of Indian mathematics, they implied a clear assessment of the superiority of European science:

> On one side, there is a perfect language, the language of Homer, approved by many centuries, by all branches of human cultural knowledge, by arts brought to high levels of perfection. On the other side, there is [in India] Tamil with innumerable dialects and that Brahmanic filth which survives to our day in the environment of the most crude superstitions. (Sedillot 1873 reprinted in Boncompagni 1964: 460)

In a similar vein Bentley (1823) also cast doubt on the chronology of India by locating Aryabhata and other Indian mathematicians several centuries later than was actually the case. He was of the opinion that Brahmins had actively fabricated evidence to locate Indian mathematicians earlier than they existed:

> We come now to notice another forgery, the *Brahma Siddhanta Sphuta*, the author of which I know. The object of this forgery was to throw Varaha Mihira, who lived about the time of Akbar, back into antiquity... Thus we see how Brahma Gupta, a person who lived long before Aryabhata and Varaha Mihira, is made to quote them, for the purpose of throwing

> them back into antiquity.... It proves most certainly that the *Brahma Siddhanta* cited, or at least a part of it, is a complete forgery, probably framed, among many other books, during the last century by a junta of Brahmins, for the purpose of carrying on a regular systematic imposition. (Bentley 1823: 151)

For the record, Aryabhata was born in AD 476 and the dates when Varahamihira, Brahmagupta and Emperor Akbar lived were AD 505, AD 598 and AD 1550 respectively. So it is safe to suggest that Bentley's hypothesis was a product of either ignorance or Eurocentric fabrication. Nevertheless Bentley's altered chronology and attitude had the effect not only of lessening the achievements of the Indian mathematics but also of making redundant any conjecture of possible transmission to Europe.

Inadequate understanding of Indian mathematics was not confined to run of the mill scholars. More recently, Smith (1923/25), an eminent historian of mathematics, claimed that, without the introduction of Western civilisation in the eighteenth and nineteenth centuries, India would have stagnated mathematically. He went on to add that: 'Not since Bhaskara (i.e., Bhaskara II, b. 1114) has she produced a single native genius in this field' (Smith 1923/25: vol. I, p. 435).

It is clear that Smith was either unaware of or ignored the works of scholars such as Whish (1835) and Warren (1825) who were among the first Westerners to acknowledge the achievements of the Kerala School. Furthermore, the alleged hiatus in astronomical and mathematical activity after Bhaskara II (b. 1114) ignores some of the influential work produced subsequently. These included Thakkura Pheru (1265–1330) of the court of the Delhi Sultanate who wrote the *Ganitasarakaumudi*; Narayana Pandit who composed an important mathematical treatise entitled the *Ganitakaumudi* in AD 1356; Mahendra Suri who composed the *Yantraraja* in 1370 which gives a table of 90 RSines with $R = 3600$; Jnanaraja who wrote the *Siddhantasundara* in AD 1503; Nityananda who wrote the *Siddhantaraja* in AD 1639; Munisvara who authored the *Siddhanta-sarvabhauma* in AD 1646; Kamalakara who wrote the *Siddhanta Tatva Viveka* in AD 1658; and Jagannatha Samrat who wrote the *Rekhaganita* in AD 1718 (a Sanskrit translation of the Arabic version of Euclid's *Elements*) and the *Siddhantasamrat* in AD 1732 (a translation of Ptolemy's *Almagest*). As shown in earlier chapters during this period astronomy and mathematics attained new heights in Kerala.

This fashion of ignoring these advances has persisted until even very recent times with no mention of the work of the Kerala School in Edwards' text (1979) on the history of the calculus nor in articles on the history of infinite series by historians of mathematics such as Abeles (1993) and Fiegenbaum (1986). A possible reason for such puzzling standards in scholarship may have been the deeply entrenched Eurocentrism that accompanied European colonisation. With this phenomenon, the assumption of European superiority became dominant over a wide range of activities, including the writing of the history of mathematics. The rise of nationalism in nineteenth-century Europe and the consequent search for the roots of European civilisation, led to an obsession with Greece and the myth of Greek culture as the cradle of all knowledge and values and Europe becoming heir to Greek learning and values. As Bernal (1987) has argued, in the 'Greek miracle' the Afro-Asiatic roots of Greece were virtually buried. What emerged as an account of the historical development of mathematical knowledge was an unreconstructed Eurocentric trajectory that ignored or devalued the contribution of the rest of the world. Rare exceptions to this skewed version of history were provided by Ebenezer Burgess (1860) and George Peacock (1849). They, respectively, wrote:

> Prof. Whitney seems to hold the opinion, that the Hindus derived their astronomy and astrology almost bodily from the Greeks.... I think he does not give the Hindus the credit due to them, and awards to the Greeks more credit than they are justly entitled to. (Burgess 1860: 387)
>
> [I]t is unnecessary to quote more examples of the names even of distinguished men who have written in favour of a hypothesis [of the Greek origin of numbers and of their transmission to India] so entirely unsupported by facts. (Peacock 1849: 420).

However, by the latter half of the twentieth century European scholars, perhaps released from the powerful influences of the colonial mentality, had started to analyse the mathematics of the Kerala School using largely secondary sources of Rajagopal and his associates (1944, 1949, 1952, 1978, 1986) and Sarasvati Amma (1963, 1979). The achievements of the Kerala School and their chronological priority over similar developments in Europe were now being aired in several Western publications (Baron 1987; Katz 1992; Calinger 1999). However, these evaluations are often accompanied by a strong defence of the European claim for the invention of the generalised calculus (Baron 1987: 65; Calinger 1999: Katz 1992: 173; 284; Plofker 2001: 283–84). For example, Baron (1987) states that:

> The fact that the Leibniz-Newton controversy hinged as much on priority in the development of certain infinite series as on the generalisation of the operational processes of integration and differentiation and their expression in terms of a specialised notation does not justify the belief that the [Keralese] development and use for numerical integration establishes a claim to the invention of the infinitesimal calculus. (p. 65)

And Calinger (1999) writes:

> Kerala mathematicians lacked a facile notation, a concept of function in trigonometry.... Did they nonetheless recognise the importance of inverse trigonometric half chords beyond computing astronomical tables and detect connections that Newton and Leibniz saw in creating two early versions of calculus? Apparently not. (p. 28)

These comparisons appear to be defending the roles of Leibniz and Newton as inventors of the generalised infinitesimal calculus. While the strength of pride in the evaluation of the achievements of these individual scientists is understandable, there is the problem of making qualitative comparison between two developments founded on different epistemological bases. It is worthwhile reiterating the point made earlier that the initial development of the calculus in seventeenth century Europe followed the paradigm of Euclidean geometry in which generalisation was important and in which the infinite was a difficult issue (Scott 1981; Katz 1992). On the other hand, from the fifteenth century onwards the Kerala mathematicians, following the different epistemology espoused by Aryabhata in the late fifth century and employing computational mathematics, were able to understand the notion of the infinitesimal and thus provide a rigorous rationale for the infinite series and the infinitesimal calculus in texts such as the *Yuktibhasa*. Further, this may be contrasted with the case of Newton who, lacking also the notion of real number, used the mysterious 'fluxions' which were only explained with the development of mathematical analysis and the clarification of the notion of 'proof' in the late nineteenth century and early twentieth century.

A modern perception appears to be that there can be no basis for mathematics apart from the Hilbertian derived Greek epistemology. Because of this perception the developments of the Kerala School of mathematics tend to be marginalised by comparing its epistemological base to the *different* Greek epistemological base. At any rate the mathematical demonstrations in the *Yuktibhasa* and other Kerala texts do not adhere to what used to be accepted ideas of a proof and so a modern pure mathematician would not have accepted them as such (and most would probably not do so even now). However, given the present debate in mathematics where the hegemony of the mathematics based on Hilbertian notions of proof is being challenged by the practical mathematics based on alternative means of justification such as computer calculation or probabilistic proofs, it is not easy to justify why acceptance of the Platonic point of view and Platonic authority *ought* to be the decisive criteria for accepting a work in mathematics. In particular, the Platonic insistence on divorce from the empirical leaves hanging in the air the question of what *logic* ought to underlie a proof as there are several different logics at hand from other cultural traditions.[1]

Another difficulty with the judgements on the value of Kerala mathematics is the limited information basis on which these are formed. As stated earlier, the current European perceptions are based on English language secondary sources that do not, as yet, give a complete understanding of the epistemology of the Kerala School. For example, the studies of many key documents such as the *Yuktibhasa* of Jyesthadeva and the *Kriyakramakari* of Sankara Variyar have only very recently become available in the European languages.

Transmission or Extrapolation? A Preliminary Discussion

The basis for establishing transmission of science is generally taken to be *direct* evidence of translations of the relevant manuscripts. The transmission of Indian mathematics and astronomy since the early centuries AD via Islamic scholars to Europe has been established by direct evidence. The transmission of Indian computational techniques was in place by at least the early seventh century for by AD 662 it had reached the Euphrates region. A general treatise on the transmission of Indian computational techniques to Europe is given by Benedict (1914) and more recently specialised literature found in Burnett (2002). Indian astronomy was transmitted westwards to Iraq, by a translation into Arabic of the *Siddhantas* around AD 760 (Subbarayappa and Sarma 1985; Saliba 1994) and into Spain, by translation into Latin of the same work in 1126. This transmission was not just westwards for there is documentary evidence of Indian mathematical manuscripts being found and translated in China, Thailand, Indonesia and other South-East Asian regions from the seventh century onwards (Subbarayappa and Sarma 1985).

In the absence of such direct evidence the following is considered to be sufficient by Neugebauer (1962) to establish transmission: *(a)* the identification of methodological similarities, *(b)* the existence of communication routes and *(c)* a suitable chronology for the transmission. Further, there is van der Waerden's 'hypothesis of a common origin' to establish the transmission of (mainly Greek) prior knowledge (van der Waerden 1983). Neugebauer uses his paradigm to establish his conjecture about the Greek origins of the astronomy contained in the *Siddhantas.* Similarly, van der Waerden uses the 'hypothesis of a common origin' to claim that Aryabhata's trigonometry was borrowed from the Greeks (van der Waerden 1976). He makes a similar claim about Bhaskara II's work on Diophantine equations, whilst offering no more than an argument based on methodological similarities. Van der Waerden is sufficiently convinced about the existence of an unknown Greek manuscript which was available to Bhaskara and his students: 'the original common source of the Hindu authors was a Greek treatise in which the whole method was explained' (van der Waerden 1976: 210). Van der Waerden concludes his discussion of the Greek origins of the works of Aryabhata and Bhaskara work by stating that '... in the history of science independent inventions are exceptions: the general rule is dependence' (van der Waerden 1976: 221).[2]

Neither Neugebauer nor van der Waerden can claim sole ownership of the transmission criteria mentioned in the foregoing paragraphs. For these have, in some form, been used by historians of science since the Renaissance. Sedillot used one aspect of the Neugebauer paradigm, communication routes, to argue for one-way transmission of astronomical ideas between Greece and India in the ancient period:

> Is it possible that the Greeks, staying in continuous contact with the Indians after Alexander's conquest, during the long reigns of the Seleucid and Ptolemy dynasties and during the existence of the Alexandrian scientific school, would remain completely estranged from the Indian science that modern scholars 'discover' today? Is it not much more likely ... that the Indians received scientific concepts from the Greeks and the Arabs, dealing with those original and creative geniuses? (Sedillot 1875: 463)

And between Europe and India in the sixteenth century:

> It is necessary to remember ... that the Portuguese took possession of Malabar coasts since the end of the fifteenth century, and it is impossible the Europeans didn't introduce through their new conquest the new ideas that propagated themselves with speed... (Sedillot 1845–1849: vol. 2, p. 465)

O'Leary uses an admixture of the Neugebauer and the van der Waerden paradigm to claim the Greek origin of Indian astronomy and mathematics. He states that:

> ... it is obvious that Indian mathematics of the period when there was a regular sea route in use between Alexandria and Ujjain were based on Alexandrian Greek teaching.... It is not necessary to prove that translations of the Greek scientists were actually made in Hindu [sic.] or Persian, it is sufficiently clear that their teaching was known and used. (O'Leary 1948: 109)

What we see from these examples is that a case for claiming the transmission of knowledge from one region to another does not necessarily rest on documentary evidence. This is a consequence of the fact that many documents from ancient and medieval times do not now exist, having perished due to variety of reasons. In these circumstances priority, communication routes and methodological similarities appear to establish a socially acceptable case for transmission from West to East. Despite these elements being in place, the case for the transmission of Kerala mathematics to Europe seems to require stronger evidence. One has merely to survey the vast majority of European accounts of the history of mathematics to date to see not a single credible mention about the possibility of this transmission. Notable exceptions to this are Brezinski

(1980) and Calinger (1999). For example, Calinger appears to give some credibility to the idea of transmission when he says:

> Donald Lach has uncovered a later technology transfer from Kerala to Europe, and it remains to be discovered whether there was a similar transmission of mathematical knowledge. (Calinger 1999: 282)

How can our conjecture of transmission possibly be proved? The tradition in Renaissance Europe was that mathematicians did not always reveal their sources or give credit to the original source of their ideas. The more well-known cases are Galileo who used sources from the Collegio Romano without acknowledgement (Wallace 1984) and Descartes against whom there is strong evidence of plagiarism from the work of the English mathematician Thomas Harriot (Fauvel and Gray 1987).

However, the activities of the monk Marin Mersenne between the early 1620s to 1648 suggest some attempt at gathering scientific information from the Orient. Mersenne was akin to being 'the secretary of the early republic of science' (Calinger 1999: 475). Mersenne corresponded with the leading Renaissance mathematicians such as Descartes, Pascal, Fermat and Roberval. Though a minim monk, Mersenne had had a Jesuit education and maintained ties with the Collegio Romano. Mersenne's correspondence reveals that he was aware of the importance of Goa and Cochin (Mersenne, in a letter from the astronomer Ismael Boulliaud to Mersenne in Rome, 1945-: vol. XIII, pp. 267–67), he also wrote of the knowledge of Brahmins and 'Indicos' (*Correspondance du P. Martin Mersenne,* 1945: vol,. XIII, pp. 518–21) and took an active interest in the work of orientalists such as Erpen—regarding Erpen he mentions his 'les livres manuscrits Arabics, Syriaques, Persiens, Turcs, Indiens en langue Malaye' (*Correspondance du P. Martin Mersenne,* 1945-: vol. II, pp. 103–15).

It is possible that between 1560 and 1650 knowledge of Indian mathematical, astronomical and calendrical techniques accumulated in Rome, and diffused to neighbouring Italian universities like Padua and Pisa and to wider regions through Cavalieri and Galileo and through visitors like James Gregory to Padua. Mersenne may have also had access to knowledge from Kerala acquired by the Jesuits in Rome and, via his well-known correspondence, could have helped in diffusing this knowledge throughout Europe. Certainly the way James Gregory acquired his Geometry after his four-year soujourn in Padua, where Galileo taught, suggests this possibility. Indeed James Gregory's own biographer states that the so-called Gregory series may have been derived from other sources:

> ... Gregory does not suggest that he is the actual author of all the theorems in this work [*Geometriae Pars Universalis*] ... We cannot judge exactly how much Gregory borrowed from other authors, because we do not know which books he may have read and to what extent he had knowledge of the unprinted work of his contemporaries. (Prag 1939: 487)

All the aforementioned issues are merely circumstantial and conjectural. To make our case for transmission we need stronger criteria. In the next chapter, in addition to the Neugebauer criteria of priority, communication routes and methodological similarities, we propose to adopt a legal standard of evidence to test the hypothesis of transmission of Kerala mathematics on the grounds of motivation, opportunity and supporting documentary evidence. However, before we do so, we will examine one transmission which has now become universal.

The Rise and Spread of Indian Numerals: A Case Study

In 1991, I concluded the first edition of my book *The Crest of the Peacock: Non-European Roots of Mathematics* with the following quotation:

> And yet if there is a single universal object, one that transcends linguistic, national and cultural barriers, and is acceptable to all and denied by none, it is our present set of numerals. From its remote beginnings in India, its gradual spread in all directions remains the great romantic episode in the history of mathematics. (Joseph 2000)

Two hundred years earlier Laplace (1749–1827) had written:

> The ingenious method of expressing every possible number using a set of ten symbols (each symbol having a place value and an absolute value)

> emerged in India. The idea seems so simple nowadays that its significance and profound importance is no longer appreciated. Its simplicity lies in the way it facilitated calculation and placed arithmetic foremost amongst useful inventions. The importance of this invention is more readily appreciated when one considers that it was beyond the two greatest men of Antiquity, Archimedes and Apollonius. (Irfah 2000)

To understand the rise and spread of the Indian numerals, it is important to distinguish between four elements that constitute the numerals: *(a)* counting with the base of 10, *(b)* the appearance of zero, *(c)* the emergence of the place-value system and *(d)* the use of the symbols for the nine digits and zero. We will examine each of these elements, but not necessarily in that order.

An important source of early information about Indian numerals comes from al-Biruni whose assessment of Indian mathematics was already referred to at the beginning of this chapter. In his book *India* composed about 1030 he wrote:

> Whilst we use letters for calculation according to their numerical value, the Indians do not use letters at all for arithmetic. And just as the shape of the letters that they use for writing is different in different regions of their country, so the numerical symbols vary. (Biruni 1030 translated by Ahmad 1999)

Historians of mathematics trace back the various symbols that al-Biruni saw to the Indian Brahmi numerals that originated around the third century BC. Figure 8.1 shows the numbers between 1 and 9. At this stage, this was not a place-value system and hence there were additional symbols for more numbers. Also, there were no special symbols for 2 and 3 with both numbers being constructed from the symbol for 1. However, the symbols for 4 to 9 have no apparent intrinsic link to the numbers that they represent, since each of the numbers is represented by a unique symbol.

Theories regarding the origin of these numerals abound. There are 'internalist' explanations that trace these symbols to the Harappan civilisation or to the later Kharosti or Brahmi alphabets or even to a system of alphabetic numerals that is no longer extant, dating back to the time of Panini. There are 'externalist' explanations that trace their

Figure 8.1

Early Indian Numerals

1	2	3	4	5	6	7	8	9
—	=	≡	+					
Brahmi numerals around 1st century A.D.								

origins to the Armenian, Greek or even Egyptian numerals. Ifrah (2000) offers both a useful survey of these conjectures and his own ingenious theory regarding the origins of Brahmi numerals:

> [T]he first nine Brahmi numerals constituted the vestiges of an old indigenous numerical notation, where the nine numerals were represented by the corresponding number of vertical lines ... To enable the numerals to be written rapidly, in order to save time, these groups of lines evolved in much the same manner as those of old Egyptian Pharonic numerals. Taking into account the kind of material that was written on in India over the centuries (tree bark or palm leaves) and the limitations of the tools used for writing (*calamus* or brush), the shape of the numerals became more and more complicated with the numerous ligatures, until the numerals no longer bore any resemblance to the original prototypes.

However, none of these explanations contains any positive evidence to back it. Whatever its origin, tracing the path of the evolution of the Brahmi numerals to our current symbols seems to be an easier matter. The next set of symbols that arose is the Gupta numerals used over large areas of the Gupta empire from early fourth century to the end of the sixth century AD. During the next five centuries the Gupta symbols evolved into the Devanagari (or Nagari) numerals shown in Figure 8.2. As the name of these numerals suggests, the symbols represent beauty and perfection—the 'writing of the gods'. These were probably the ones that al-Biruni referred to when he wrote: 'What we use for numerals is a selection of the best and most regular figures in India.' And these 'most regular figures', including a symbol for zero, by al-Biruni's time had been transmitted to the Islamic world.

Figure 8.2

Indian Numerals around the 11th Century

1	2	3	4	5	6	7	8	9	0
१	२	३	४	५	६	७	८	९	०
Nagari numerals around 11th century A.D.									

Source: Joseph (2000: 216–19).

Two aspects of the new Indian number system are noteworthy: the existence of a symbol for zero and the operation of a place-value system with the numerals standing for different values depending on their position relative to the other numerals. Although our place-value system is a direct descendant of the Indian system, it should be noted that the Indians were not the first to develop such a system.

There are historical records of three other number systems which were based on the positional principle. Predating all other systems was the Mesopotamian system which must have evolved around the third millennium BC. A base-60 scale was employed, with a simple collection of the correct number of symbols, to write numbers less than 60. But it was imperfectly developed being partly additive and partly positional for within the base-60 a decimal system was used. Also, the absence of a symbol for zero until the early Hellenistic period limited the usefulness of the system for both computational and representational purpose.[3]

The Chinese rod numeral system was essentially a decimal base system. The numbers 1, 2, …, 9 are represented by rods whose orientation and location determine the place value of the number represented and whose colour show whether the quantity was positive or negative. In terms of computation, the representation of zero by a blank space posed no problem for, unlike the Babylonian system, the blank was itself a numeral.[4]

The third positional system was the Mayan, essentially a vigesimal (base 20) system incorporating a symbol for zero which was recognised as a numeral in its own right. It had, however, a serious irregularity since its units were 1, 20, 18×20, 18×20^2, 18×20^3, … and so on. This anomaly reduces its efficiency in arithmetical calculation. For example, one of the most useful facilities of our number system is the ability to multiply a given number by 10 by adding a zero to it. An addition of a Mayan zero to the end of a number would not in general multiply the number by 20 because of the mixed base system employed which meant that corresponding to our units, tens, hundreds, thousands, ten thousands, etc, the Mayans had units, 20s (18×20^2), (18×20^3) …, etc.[5]

The oldest dated Indian document containing a number written in the place-value format that we use today is a legal document dated 346 (or 594 AD in our calendar) from the Bharukachcha (or Broach) region in Gujarat. This would indicate that a place-value system was in use in India from at least the end of the sixth century.[6] The first written record, both dated and not disputed, is the inscription at Gwalior dated 'Samvat 933' in the Vikrama calendar which translates to a date in our calendar of AD 876. In the same inscription, the numbers 50 and 270 are recorded with a small circle appearing at the appropriate positional place for zero.

An interesting question arises as to why the Indians were the first to develop such an innovative system. Explanations have ranged from the Indian system being an adaptation of the Mesopotamian base-60 place-value system to a base-10 system as a result of the borrowing of the idea from the Chinese positional rod numeral system which, according to Lay Lam Yong had the '... three essential features of our numeral notation system: *(a)* nine signs and the concept of zero, *(b)* a place value system and *(c)* a decimal base'.[7]

A more plausible hypothesis, at least according to the author, is that the idea of the place-value system with a zero gradually evolved internally in India over centuries. I have argued elsewhere (Joseph 2000) that Indian fascination with large numbers lies at the root of the development of this innovative Indian system of numerals.

Fascination with large numbers has been an abiding characteristic of Indian civilisation.[8] As early as 800 BC, there appeared names for powers of 10 up to 62. The importance of these number-names in the evolution of the decimal place-value notation cannot be exaggerated.[9] The word-numeral system was the logical outcome of proceeding by multiples of 10. In an early word-numeral system going back over 2,500 years ago, 60,799 is denoted by the Sanskrit word, *sastim* (60), *sahsara* (thousand), *sapta* (seven) *satani* (hundred), *navatim* (nine ten times) and *nava* (nine).

Such a system presupposes a scientifically based vocabulary of number names in which the principles of addition, subtraction and multiplication are used. The system required: *(a)* the naming of the first nine digits (*eka, dvi, tri, catur, pancha, sat, sapta, asta, nava*); *(b)* a second group of nine numbers obtained by multiplying each of the nine digits in *(a)* by ten (*dasa, vimsat, trimsat, catvarimsat, panchasat, sasti, saptati, astiti, navati*); and finally *(c)* a third group of numbers which are increasing integral powers of 10, starting with 10^2 (*sata, sahasara, ayut, niyuta, prayuta, arbuda, nyarbuda, samudra, madhya, anta, parardha,* .....). In forming the words of the second and third groups of numbers, the multiplicative principle is used, as for the example quoted, 60,000 is *sastim-sahasari.* The additive principle is employed when the numbers from the first and second group are used, for example, 27 is *sapta-vimsati*. The subtractive principle may apply occasionally and in a limited way, as for example, *ekanna-catvarimsat* indicates $20 - 1 = 19$, where *ekanna* means 'one less'.

The discussion so far of the different positional systems highlights two important points. First, a place-value system can and did exist without any symbol for zero. But the zero symbol as part of the numerical system never existed and could not have come into being without place value. Second, the relative strength of the Indian number system is inextricably tied up with the Indian concept of zero. It would, therefore, be useful to examine the origins and use of zero in Indian mathematics.

The Sanskrit word for zero, *sunya*, means 'void or empty'.[10] Its derivative, *sunyata*, is the Buddhist doctrine of Emptiness, being the spiritual practice of emptying the mind of all impressions. This is a course of action prescribed in a wide range of creative endeavours. For example, the practice of *sunyata* is recommended in writing poetry, composing a piece of music, in producing a painting or any activity that came out of the mind of the artist. An architect is told in the *silpa sutras* (traditional manuals of architecture) that designing a building involves the organisation of empty space, for 'it is not the walls which make a building but the empty spaces created by the walls'. The whole process of creation is well described in the following quotation from a Tantric Buddhist text (*Havraja Tantra*):

> First the realisation of the void (*sunya*),
> Second the seed in which all is concentrated,

> Third the physical manifestation,
> Fourth one should implant the syllable. (Datta 1927)

The mathematical correspondence was soon established. 'Just as emptiness of space is a necessary condition for the appearance of any object, the number zero being no number at all is the condition for the existence of all numbers.'

It was soon recognised that the *sunya* denoted notational place (place holder) as well as the 'void' or absence of numerical value in a particular notational place. Consequently, all numerical quantities, however great they may be, can be represented with just 10 symbols. A twelfth-century text (*Manasollasa*) states: 'Basically, there are only nine digits, starting from "one" and going "nine". By the adding zeros these are raised successively to tens, hundreds and beyond' (Arundhati 1994).
And in a commentary on Patanjali's *Yogasutra* there appears in the seventh century the following analogy:

> Just as the same sign is called a hundred in the 'hundreds' place, ten in the 'tens' place and one in the 'units' place, so is one and the same woman referred to (differently) as mother , daughter or sister.

The earliest mention of a symbol for zero occurs in the *Chandasutra* of Pingala (fl. third century BC) which discusses a method for calculating the number of arrangements of long and short syllables in a metre containing a certain number of syllables (that is, the number of combinations of two items from a total of *n* items, repetitions being allowed).[11] It began as a dot (*bindu*), found on inscriptions both in India and in Cambodia and Sumatra around the seventh and eighth century and then became a circle (*chidra* or *randra* meaning a hole). The association between zero and its symbol had become well known around the early centuries of the Christian era, as the quotation shows: 'The stars shone like zero dots (*sunya-bindu*) ... scattered in the sky on a blue rug, with the Creator reckoning the total using a bit of the moon as chalk.' (Vasavadatta, c. AD 400) (see Bag and Sharma 2003).

Sanskrit texts on mathematics/astronomy from the time of Brahmagupta (b. 598) usually contain a section called '*sunya-ganita*' or computations involving zero.[12] While the discussion in the arithmetical texts (*patiganita*) is limited only to the addition, subtraction and multiplication with zero, the treatment in algebra texts (*bijaganita*) cover such questions as the effect of zero on the positive and negative signs, division with zero and more particularly the relation between zero and infinity (*ananta*).

Take as an example, Brahmagupta's *Brahmasphuta-Siddhanta*. In it he treats the zero as a separate entity from the positive (*dhana*) and negative (*rhna*) quantities, implying that *sunya* is neither positive nor negative but forms the boundary line between the two, but forms the boundary line between the two, being the sum of two equal but opposite (i.e., positive and negative) quantities. He states that a number, whether positive or negative, remains unchanged when zero is added to or subtracted from it. In multiplication with zero, the product is zero. Likewise the square and square root of zero is zero. But when a number is divided by zero, the answer is an undefined quantity 'that which has that zero as the denominator'.

The story of the spread of the Indian numerals is a fascinating one. It took place at different times, in different directions, at different pace and for different reasons. In China, with the spread of Buddhism, there was an exchange of knowledge. Buddhist pilgrims from China visiting India carried back large quantities of manuscripts with them; Indian scholars who went to China were engaged in translating Buddhist texts into Chinese. Although existing records only mention religious and philosophical texts, it is possible that ideas or even texts relating to Indian systems of numeration, mathematics and astronomy were brought into China. We know, for example, from the catalogue of the Sui dynasty, completed in AD 610, that there were Chinese translations of Indian works on astronomy and mathematics. These translations are no longer extant, but the fact that they existed showed that by the end of the sixth century AD, the Chinese may have acquired some knowledge of Indian mathematics and astronomy. Of greater relevance were records from the seventh century onwards of the Tang dynasty, from which we gather that Indian astronomers were employed at the Astronomical Board of Chang-Nan to teach the principles of Indian astronomy and calendar. An Indian named Gautama Siddhartha was reported to have constructed a calendar along the lines indicated in the Indian *Siddhantas*. This calendar contains sections on Indian numerals and arithmetical operations and sine tables at intervals of 3°54′ for a radius of 3,438 units, which are precisely the values given in the Indian astronomical texts. There survives a block print text which contains Indian numerals, including the use of a dot to indicate zero.[13] Yet, despite the influence of Indian astronomy, the impact of Indian numerals was hardly felt in China. As Sarma (2009: 213) writes:

> Strangely enough, neither the dot for zero found in this calendar prepared by Gautama Siddhārtha, nor the dot and circle for zero found in inscriptions in what Needham refers to as the 'southern zone of the culture of the Chinese' in the seventh century seem to have had any impact in China proper, because the Chinese did not start using the zero symbol until the mid-thirteenth century when it appears for the first time in the work of Qin Jiushao. It is rather intriguing why the Chinese took such a long time (nearly six centuries) to use the zero symbol which was brought up to their doorstep, so to speak. Here is a case of clear transmission having no impact in the receiving culture. The same appears to be the case with the other elements of Indian mathematics and astronomy introduced into China in the Tang period.

The spread of the Indian numerals westwards is a different story.[14] The Islamic world figured as leading actors in this drama. Indian numerals probably arrived at Baghdad with the diplomatic mission from Sindh to the court of second Abbasid Caliph al-Mansur (753–774).[15] However, there is evidence that knowledge of the Indian decimal place-value system reached the Middle East earlier, not through this visit of this mission but at least a century earlier. In AD 662 the Nestorian Bishop Severus Sebokht sang praises to the Indian decimal numbers in a passage that is frequently quoted.[16]

In about AD 820 al-Khwarizmi wrote his famous *Kitab al-jam'wal tafriq bi hisab al-Hind* (or 'Arithmetic'), the first Arab text to deal with the new numerals. This was one of the three works that he authored, the others being in algebra and mathematical astronomy. In the work on arithmetic al-Khwarizmi explained the operations of addition, subtraction, multiplication, division and the extraction of square roots according to the Indian system. There were other introductions to Indian arithmetic that followed al-Khwarizmi's work. But the availability of his work in Spain in the twelfth century led to it being translated into Latin, under the title *Liber algorismi de numero Indorum*. Soon the value of al-Khwarizmi's book was recognised and there appeared more commentaries in Latin on his book. What is interesting is that while the symbol for zero which was a small circle remained unchanged in subsequent evolution of the numerals in Europe, the digits 1 to 9 gradually went through changes even to the extent of the Islamic world developing two different forms (the Western forms or the Gubar numerals and the Eastern forms which is the direct ancestor of what is described as the 'Arabic numerals' or what the Indian constitution more accurately terms in Article 343.1 as 'the international form of Indian numerals'[17]). Al-Khwarizmi was at pains to point out the usefulness of a place-value system incorporating zero, particularly for writing large numbers. Texts on Indian reckoning continued to be written and by the end of the eleventh century, this method of representation and computation was widespread from the borders of Central Asia to the southern reaches of the Islamic world in North Africa and Egypt.

In the transmission of Indian numerals to Europe, as with almost all knowledge from the Islamic world, Spain and (to a lesser extent) Sicily played the role of intermediaries, being the areas in Europe which had been under Muslim rule for many years. Documents from Spain and coins from Sicily show the spread and the slow evolution of the numerals, with a landmark for its spread being its appearance in an influential mathematical text of medieval Europe, *Liber Abaci*, written by Fibonacci (1170–1250) who learnt to work with Indian numerals during his extensive travels in North Africa, Egypt, Syria and Sicily. And the spread westwards continued slowly, displacing Roman numerals, and eventually, once the contest between the *abacists* (those in favour of the use of abacus or some mechanical device for calculation) and the *algorists* (those who favoured the use of the new numerals) had been won by the latter, it was only a matter of time before the final triumph of the new numerals occurred with bankers, traders and merchants adopting the system for their daily calculations.

There are other areas of westward transmission of Indian mathematics worthy of study apart from the number system alone. Case studies can be conducted along similar lines as mentioned in the foregoing paragraphs on *(a)* the spread of Indian trigonometry, especially the use of the sine function and *(b)* the solutions of equations in general, and of indeterminate equations in particular. In this context, studies of simple procedures such as the Rule of Three (*Trairasika*) or extraction of square or cube roots or use of the results of right-angled triangle and their possible transmission could also be useful exercises.[18]

Notes

1. These ideas owe a considerable debt to Paul Ernest. For further details, see Ernest (2009: 181–202).
2. There is a long history of attributing Greek origins to Indian mathematics. For example, Cantor (1880–1908), having expressed admiration for Indian algebra, took every opportunity to emphasise its Greek antecedents. For further details, see Cantor (1892: vol. II, pp. 562, 584).
3. For a discussion of the Mesopotamian system, see Joseph (2000: 96–103).
4. For a discussion of the Chinese rod numeral system, see Joseph (2000: 140–48).
5. For a discussion of the Mayan numeral system, see Joseph (2000: 50–51).
6. However, there are some historians who claim that this document is a later forgery. There are other written evidence in the form of documents and inscriptions found during the eighth and ninth centuries mentioned by Ifrah (2000).
7. See Lam Lay Yong (1986, 1987).
8. In *Ramayana*, the great Hindu epic about 2,500 years old, there is a description of two armies facing each other. The size of the larger army led by Rama is given as follows in a seventeenth-century Malayalam translation of the epic by Kottayam Kerala Varma Thampuram:

   Hundred hundred thousands make a Crore
   Hundred thousand crores make a Sankhu
   Hundred thousand sankhus made a Maha-sankhu
   Hundred thousand maha-sankhus made a Vriundam
   Hundred thousand vriundam made a Maha-vriundam
   Hundred thousand maha-vriundams make a Padmam
   Hundred thousand padmams make a Maha-padmam
   Hundred thousand maha-padmams make a Kharvam
   Hundred thousand kharvams make a Maha-kharvam
   Hundred thousand maha-kharvams make a Samudra
   Hundred thousand samudras make a Maha-yougham
   There is no end to the naming of numbers.

9. An early recognition of the meaning of the place-value notation is found in the following simile from *Vyasabhasya* (a commentary) on the *Yogasutra* (III.13), dating around the beginning of the Christian Era.

   Yathaika rekha satasthan satam dasasthane dasaika caikasthane tatha caikatve 'pi stri mata cocyate duhita ca svasa cati/

   ['Just as a stroke denotes 100 in the "hundreds" place, 10 in the "tens" place and 1 in the "units" place, even so one and the same woman is called mother, daughter and sister (by different persons).'] (Joseph 2000: 241, 311–12)

10. The word *sunya* is derived from *suna* which is the past participle of svi, 'to grow'. In one of the early Vedas, *Rigveda*, occurs another meaning: the sense of 'lack or deficiency'.

It is possible that the two different words were fused to give '*sunya*' a single sense of 'absence of emptiness' with the potential for growth. The following etymology for zero tracing it back to *sunya* has been suggested:

Sunya → *Sifr* (Arabic) → *Zephizum* (Latin) → *Cipher or Zero*

11. In showing how to calculate the number of permutations of a verse containing two sets of syllables—one short and the other long—Pingala outlines a procedure in which the steps in the calculation are labelled as *dvih* (two) and others as *sunya* (zero). Here the symbols are merely used as markers for showing two different kinds of operations. The symbol of two indicates a place where there is an even number (and so halving is possible) and that of zero shows the stage where there is an odd number (so absence of halving and multiplication by 2 performed). The fact that that the two symbols were used as markers would indicate that a widely-used symbol for *sunya* existed, although we don't know what it is. A symbol would presuppose a concept. It would indicate in this case the absence of an operation. For further details, see Sarma (2003), to whom this endnote is indebted.
12. Operations with zero have been mentioned in Chapter 6 where we detected a hint of an understanding of the concept of a limiting process.
13. For further details and reference, see Sarma (2009: 215–16). I am greatly indebted to Professor Sarma for some of the information on this section.
14. There is earlier evidence of the use of Indian system of numeration in South-East Asia in areas covered by present-day countries such as Malaysia, Cambodia and Indonesia, all of whom were under the cultural influence of India. Also, as early as AD 662, a Syrian bishop, Severus Sebokt, comments on the Indians carrying out computations by means of nine signs by methods which 'surpass description'. For further details, see Joseph (2000).
15. The arrival of this mission is reported in a number of places. These include al-Biruni, al-Adami and Abu-Masher who stated that they had acquired their knowledge regarding star-cycles and other astronomical phenomena and methods from an Indian astronomer who was well versed in the sciences. The name of this astronomer has been variously reported as 'Kankaraf', 'Kankah', 'Cancah', 'Kenker', or 'Kanaka', and he has become almost a mythical figure in terms of his vast knowledge and scholarship. Sarma (2008) has an interesting theory: 'Perhaps the real word was not "Kanaka" but "Gaṇaka" - not a proper name but a generic name for the astronomer. Probably this word referred not to one particular astronomer who visited Baghdad in 771, but collectively to all Hindu astronomers or learned people who visited Baghdad about this time. This would explain the diverse qualities attributed to this *Gaṇaka* who may have, in reality, represented different persons.'
16. The passage reads:

    I will omit all discussion of the science of the Indians, a people not the same as the Syrians; of their subtle discoveries in astronomy, discoveries that are more ingenious than those of the Greeks and the Babylonians; and of their valuable methods of calculation which surpass description. I wish only to say that this computation is done by means of nine signs. If those who believe, because they

> speak Greek, that they have arrived at the limits of science, [would read the earlier texts], they would perhaps be convinced, even if a little late in the day, that there are others also who know something of value. (Sarma 2009: 214)

17. Sarma (2009: 229) has put the dilemma regarding the nomenclature clearly:

    > This is no doubt an elegant solution for the nomenclature of the numerals of the form 1, 2, 3, and so on. But how does one designate unambiguously the numerals associated with the Arabic-Persian script, such as ١٢٣? Again, when the Indian numerals were transmitted to Europe, they were first known as the 'Indian' numerals. One wonders when Europe started calling them 'Arabic' numerals and why.

18. See for example, the case of the 'Rule of Three' in Sarma (2009: 216–19).

9

Exploring Transmissions: A Case Study of Kerala Mathematics*

Introduction: The Legal Case for Transmission

A refinement and extension of the earlier discussion of transmission in the last chapter may be achieved with the introduction of the 'legal' criterion. To the Neugebauer criteria (that is, identification of methodological similarities, establishing the existence of communication routes and constructing a plausible chronology for the transmission) we add the legal standard of proof—that is, establishing the existence of motivation, opportunity and 'incriminating' evidence, both of a circumstantial and documentary nature. We are aware that a general motivation for the import of Indian mathematical knowledge to Europe was the growing interest in better computing techniques needed for more accuracy in navigation, calendar making and practical mathematics.[1] Given that Indian mathematics had already discovered some of the

*This chapter is based on papers presented at a workshop held in Kovalam (Kerala) and published as an edited volume of the Proceedings. Three papers in particular are referred to, namely Almeida and Joseph (2009a), Baldini (2009) and Bala (2009). The author is grateful to Ugo Baldini and Arun Bala for having presented different perspectives on the transmission issue so clearly.

advanced mathematical techniques at the time Europeans were searching for them, they would have been motivated to learn these should the opportunity arise.

The opportunity presented itself to the Europeans as a result of the route they discovered to reach India in 1498 by rounding the Cape of Good Hope. The Portuguese landed at Calicut and Cochin in Kerala, the heartland for the development of the new mathematical ideas that we explored earlier. Even if initially the Portuguese were not been able to understand the Indian mathematical literature they would have been in a much better position to do so after the arrival of Jesuit missionaries, some of whom were trained in mathematical, navigational and astronomical sciences.

In this chapter we examine the findings of a research project which uses an extension of the objective criteria outlined earlier,[2] to investigate our conjecture regarding possible transmission of Kerala mathematics to Europe through the Jesuit conduit. Specifically, we seek in the first instance to find *direct* documentary evidence of transmission and, in the event of the failure to discover such evidence, we will also examine the veracity of the existing belief that there was no transmission of the Kerala mathematics to Europe using the legal criteria. Or to borrow an analogy from mathematical statistics, we aim to test the *status quo* hypothesis (that is, the null hypothesis) that there was no transmission of (influence of) the Kerala mathematics to (in) Europe *against* our conjectured hypothesis (that is, the alternative hypothesis) that there was transmission and influence.[3] The aim is to study materials that were identified as providing supporting evidence for either of the hypotheses. The evidence sought would not only be of the 'smoking gun' circumstantial type but also direct evidence of the process or act of transmission having occurred as identified in documents.

The Data

The manuscripts and other materials that were identified in the research project for investigation together with their locations and the associated justifications for their inclusion are given in the following paragraphs. Materials were initially identified by the project members and copies of those that appeared to warrant specialist translation and study were obtained and sent to our research assistants who had the necessary linguistic and technical skills. Additionally, one of them (Goncalves) spent an extended time in Portugal to map out the existence of relevant materials in the un-catalogued documents in the archives at the University of Coimbra; another (Delire) undertook a study of the catalogue of Oriental manuscripts in Leiden University.

A. Letters and Reports of Jesuit Missionaries in Kerala: 1540–1650

The arrival of Francis Xavier in Goa in 1540 heralded a continuous presence of the Jesuit missionaries in the Malabar (the region surrounding Kerala) till 1670. Some of the Jesuits arriving after 1578 appeared to have the objective of gathering information from India.[4] These Jesuits trained either by Clavius or Grienberger (Clavius' successor as Mathematics Professor at the Collegio Romano) were sent to India. Most notable of these were Matteo Ricci, Johann Schreck and Antonio Rubino. Ricci had specialist knowledge of mathematics, cosmography, astronomy and navigation; the Jesuit historian Henri Bernard states that Ricci '…had been requested to apply himself to the scientific study of this new and imperfectly known country, in order to document his illustrious contemporary, Father Maffei, the "Titus Livius" of Portuguese explorations' (Bernard 1973). Rubino had studied with the French mathematician Viete, well known for his work in algebra and geometry. At some point in their stay in India these Jesuits went to the Malabar region, including the city of Cochin, the epicentre of developments in Kerala mathematics.[5]

Indeed, from an earlier study, we have identified the following items of circumstantial evidence to support transmission:

1. These Jesuit missionaries were interested in arithmetic, astronomy and timekeeping of the Malabar region.[6]
2. These Jesuit missionaries were able to appreciate this knowledge by their learning of the vernacular languages such as Malayalam and Tamil (Wicki 1948-: Vol. XIV, p. 425; Vol. XV, p. 34).

3. Local sciences such as astrology or *jyotisa* were included in the curriculum of the Jesuit colleges in the Malabar Coast (Wicki 1948-: Vol. III, p. 307).
4. There were descriptions of the sciences and the mechanical arts of the Malabar region sent to Rome.[7]
5. Ricci's enquiries in 1580 about Indian calendrical science (Wicki 1948-: Vol XII, p. 474).
6. Rubino's report in 1610 about the errors in European tables based on inferences from local calendrical knowledge (Baldini 1992: 214).
7. The letter from Schreck, in 1618, of astronomical observations intended for the benefit of Kepler (Iannacone 1998: 58).
8. Rubino stated in a letter to Grienberger about the Malabar Brahmins who 'he wrote, "are devoted to study of the movements and aspects of the planets and stars, particularly of twenty seven by which they govern and rule." He tried to learn their secret of predicting "the hour and minute of eclipses of the sun and moon," but was unsuccessful because they shared this knowledge only with relatives and in secret' (D'Elia 1960: 15).

Thus it was relevant to study the following mass of materials to determine whether or not there was direct evidence of transmission:

1. Archivum Romanicum Societate Iesu (ARSI), Rome. A study of mainly manuscript letters and reports from Jesuit missionaries to their headquarters in Rome.
2. Gregorian University Archives, Rome. A study of manuscript correspondence of scientist Jesuits [amongst them Rubino, Ricci, Schreck] to Clavius and Grienberger.
3. University of Coimbra archives, Coimbra. An investigation to identify any Jesuit correspondence and materials and to examine the works by the Jesuit mathematician Borri. He was the only Jesuit missionary who went to the Malabar and returned. He specialised in astrology.
4. Ajuda library, Lisbon. A study of manuscript correspondence of the earlier Jesuit missionaries to the Malabar up to 1568.

B. Works of Jesuits in Rome Who may Have Come across Kerala Mathematics

There appears to be a similarity in the approach to certain result in the proto-typical calculus in the *Yuktibhasa* and the approach to calculus adopted by some Renaissance mathematicians. Additionally some results of Indian mathematics of 500 years earlier—those of Bhaskara II—were rediscovered in the Renaissance by Fermat and Wallis.[8]

If the Jesuits were the conduit of the conjectured transmission then it would require Jesuit mathematicians of some ability to interpret the works of the Kerala mathematicians. The most prominent of the Jesuit mathematicians of the period in question were the two mentioned previously, Christopher Clavius[9] and Christopher Grienberger.[10] Additionally, we know from our earlier studies[11] that scholarly Jesuit missionaries in the Malabar such as Ricci and Rubino were in correspondence with Clavius and Grienberger.

The materials studied were as follows:

1. Works and correspondence of Christopher Grienberger. The correspondence was identified from the edited correspondence of Clavius (Baldini and Napolitani 1992). The mathematical works we examined were his unstudied manuscripts GES 874 and GES 600 in the Biblioteca Nazionale, Rome.
2. Works and correspondence of Christopher Clavius. The correspondence was identified in the edited correspondence as mentioned earlier paragraphs and the work identified was the one work which may have been susceptible to influence from Indian sources: *Theodosii Tripolitae Sphaericirum Libri III*, located in the Univeristaria Alessandrina in Rome. This deals with the calculation of the trigonometric ratios which we posited to have some connection with the Kerala mathematics.

C. Indian Scientific MSS Located in Libraries in Europe

Our studies prior to the present project included an examination of the correspondence of Renaissance mathematicians organised by Marin

Mersenne (*Correspondance du P. Marin Mersenne* 1945-). The minim monk Marin Mersenne, through his correspondence with the leading scientists and mathematicians of the early seventeenth century, was an important conduit for transmission of knowledge. In this correspondence there is mention of Brahmins (*Correspondance du P. Marin Mersenne* 1945- : Vol. XIII, pp. 518–21)[12] and of the orientalists Gaulmin (*Correspondance du P. Marin Mersenne* 1945- : Vol. XIII, pp. 518–21), Erpen and his 'les livres manuscrits Arabics, Syriaques, Persiens, Turcs, Indiens en langue Malaye' (*Correspondance du P. Marin Mersenne* 1945- : Vol. II : pp. 103–115) Golius and Drusius in the University of Leyden (*Correspondance du P. Marin Mersenne* 1945- : Vol. II, p. 155).[13]

Additionally, our prior studies included an initial survey of catalogues in the Vatican library. These indicated the presence of a large number of palm leaf manuscripts in Malayalam (the language of Kerala) and Tamil (the language of the neighbouring state, Tamil Nadu). Thus we undertook an examination of manuscripts whose existence we discovered in our earlier studies. These consisted of:

1. Malayalam and Tamil palm leaf manuscripts in the Vatican library.
2. Oriental manuscripts in Leiden University.

Results of the Investigation

The materials identified in the foregoing paragraphs were studied in-depth. Being primary archival research the net that we cast over the materials was fine and therefore necessarily captured large amounts of information not directly relevant to the aims of the research. For example, the further examination of Jesuit correspondence at the ARSI revealed only material related matters related to the missions (conversions, finance, the establishment and administration of the colleges, and so on). Similarly the material at the Ajuda library did not reveal any evidence of scientific information gathering and reporting from the Jesuit missionaries in the Malabar. Examples of letters from the Ajuda (Ajuda Archives, Rome) are shown as follows:

Folio 49-IV-50; Letter No. 192. Carta do Irmao Luis Frois para casas e collegios da companhia de Europa escrita em Goa ao derradeiro de nouembro de 1557 [fl 98] [- fl 108v: 20 pages. About Brahmans, moors, etc]

Folio 49-IV-50; Letter No. 195. Copia de outro do Irmao Luis Frois do collegio de Goa a 14 de nouembro de 1559. [fl 120v] [-fl 131v: 22 pages. About Brahmans, moors, etc]

Folio 49-IV-50; Letter No. 264. Parte dalguas cousas de o Irmao Luis Frois escreuo da India ao Irmao Volgango germano companhia de Jesus no collegio de Coimbra a 30 de nouembro de 1560. [fl 333v] [-fl 335v: 6 pages. About Brahmans, moors, etc]

We report that the evidence from the examination of the Jesuit documents and correspondence merely confirmed the findings of our prior study that there was strong motivation on the part of the Jesuit missionaries to acquire the science of the Malabar.[14] However, the evidence from the trawl of materials studied so far does not support the contention that the Jesuits acquired any of the manuscripts that contained the Kerala mathematics and neither was there any direct evidence that they acquired knowledge of these results from a third party.

The correspondence between Clavius and Grienberger on the one hand and the Renaissance mathematicians in contact with them [notably Adrian van Roomen] on the other hand suggest that they were working in an epistemology which seemed uninfluenced by the Kerala mathematics. However, influences of the earlier Indian mathematics notably from that of Aryabhata were detected. In this context it should be pointed out that Clavius was essentially re-writing the earlier work on trigonometric tables by Regiomontanus and thus it will be truer to say that the latter rather than Clavius was influenced by Aryabhata. For in Regiomontanus's *Compostio Tabularum suinuum Rectorum,* published in 1541 he, in the manner of Ptolemy, calculates the sines of 90°, 45°, 60°, 30° and 15° from appropriate triangles, and then uses the 'Pythagorean' theorem to calculate sine 75°. He then uses the formula 'For a random arc in a quadrant, the sine is the middle proportional [geometric mean] between half the radius and the versed sine of double the arc' which is equivalent to Aryabhata's earlier (c. AD 499) *kramatkramajya* rule (Shukla and Sarma 1976). This finding is in accordance with Knobloch's work[15] on the Arab influence on Clavius's work as we note that Arab mathematics was itself influenced by developments in India.[16] Having said that, a more detailed comparative study needs to be made between the early European work on the construction of trigonometric tables and the Kerala work on similar tables.

Grienberger's hitherto unstudied manuscript works GES 674 and GES600 were examined in detail. It revealed a lengthy work on the computation of trigonometric tables graduated in degrees with an intended accuracy of 18 places of decimals. The novel methods and algorithms used in the lengthy calculations were deciphered with some difficulty with the assistance of the Research Assistants. It revealed that Grienberger's methods were outstanding and novel, utilising methods from many sources, but there was no evidence on influence from Kerala mathematics (although the arithmetic used is Indo-Arabic). We (and, implicitly, several others) had conjectured that the initial value of sine of 1 minute [correct to 22 places of decimals] used by Grienberger to generate his tables was calculated using the infinite series for sine that had been discovered earlier by the Kerala mathematicians. However, our analyses made it clear that the infinite series of trigonometric ratios is impracticable for the construction of trigonometric tables graduated in degrees. Nevertheless the studies of Clavius and Grienberger, especially the latter, on the construction of trigonometric tables will be of interest to historians of mathematics and work is ongoing to publish these findings.

A thorough study of a set of uncatalogued Malayalam and Tamil palm leaf manuscripts at the Vatican library was undertaken by Dr Raghava Varier, a research scholar of Malayalam and Tamil literature. No evidence of Kerala mathematics or astronomy was found in these manuscripts. The manuscripts were mainly works on lexicography and catechisms with one or two works on medicine. A full report on the archival work undertaken by Dr Varier is contained in one of the papers ('Uncatalogued Malayalam Manuscripts in Europe: A Report of Work Carried out in Two Libraries in Rome') to be found in the volume of the Workshop Proceedings (Joseph 2009).

Amongst the oriental manuscripts at Leiden there were several that contained works of early Indian mathematics. For example, Or. 2361. Sanskrit, paper, 42 ff. *Mahasiddhanta* by Aryabhata II (fl. AD 950) in 18 *adhyayas*. However, the history of acquisition of these manuscripts was missing and thus nothing could be inferred about the possible transmission implications. Furthermore, the mathematical works related to a period of Indian mathematics prior to the emergence of Kerala School.

Certain Tentative Conclusions

The painstaking trawl of the mass of manuscript and other materials, mentioned earlier in this chapter, has yielded no direct evidence of the conjectured transmission. Therefore, we have to report that on the basis of the evidence of the documents studied so far that the evidence supports the null hypothesis formulated earlier. Thus, on the basis of this set of evidence, the European Renaissance developments of prototypical calculus may well have been independent of the developments in that subject in Kerala some centuries earlier.

This is only a provisional conclusion as it is by no means the case that all of the material that is required to be studied to reach a definitive conclusion has been studied. There may be materials available amongst the mass of uncatalogued documents in Portugal or in private libraries in Italy. Both these countries suffered upheavals in which library contents were dispersed—in Portugal, the suppression of the Jesuits in 1750 by the Marquis of Pombal and in Italy, the effects of the thirty years war in the early seventeenth century. Additionally, the unstudied works of other Renaissance mathematicians (such as Toricelli and Cavalieri) could be relevant to the examination of our hypothesis.

Also, the possibility of oral transmission does exist. However, weighed against this possibility is the fact, as far as can be determined, that only Cristovão Borri of the scientifically accomplished Jesuit missionaries eventually returned from Malabar to Europe. And we have found no evidence of any information from him being available in his publications or reported on by the scholars with whom he was in contact. Nevertheless, it may be the case that the oral conduit may have been an Arab one with the sea route to Basra from Kerala being still active in the period in question. Given the new information, some of it oral, on the crucial role of the Arabs in the creation of the Copernican Revolution, it may be useful to pursue this line of thought in any future research. We report in a later section Dr Arun Bala's conjectures regarding how skill-based knowledge can be transmitted without documentation.

In attempting to study transmissions, it is also important to examine in greater detail the context and the motives of the Jesuit missionaries who were sent to India and their mode of communication with one another. The primary motive was of course evangelical but to achieve this different kinds of strategies were adopted. In their mission to India and China, it was recognised early on that they sent missionaries who were both well informed on the sciences and technology of the day and could debate with local scientists that they came across. It is unlikely, given the nature of the relationship between themselves and the host societies (who were often described as pagans and unbelievers),[17] that they would openly concede or acknowledge the superiority of indigenous scientific knowledge. The position of the Jesuits was during the period under study quite precarious, as the Galileo Episode indicated. In any case, within Europe of that period, the intellectual climate did not favour any explicit acknowledgement of the debt owed to the work of others, unless they were ancient Greeks who were perceived as the fount of knowledge from which European history and thought emerged.

From the beginning of the Jesuit missionary-scientific enterprise, great importance was placed on the Jesuits in far-flung places for keeping in touch with others. Francis Xavier, the first missionary in the East, urged the members to pay particular attention to the composition of the letters, both public and private, making the former suited to the needs. Ground rules were laid down on the manner in which communication was to be conducted.[18] It would seem unlikely that an organisation which had enemies in Rome and was perceived at times as sailing close to wind, particularly during the time of the Galileo Episode, would admit either in the letters or elsewhere that they were recipients of 'superior' knowledge in science from 'pagans and unbelievers'. So the absence of documentary evidence of transmission should not come as such a great surprise.

At any rate what we do know from the project study is that the null hypothesis of no direct transmission is sustained by the evidence gathered. If in the final analysis—by extended archival work or from other sources over many years—this null hypothesis is confirmed, then it will be the first major case of scientific development in the post-Ancient era that has remained localised in its place of origin, and that despite the existence of a direct corridor of communication to Renaissance Europe.

What reasons might be put forward for this if in the final analysis the Kerala mathematics remained localised and un-influential despite the conditions for its ease of transmission being present? The combination of Jesuits' colonial superiority and insensitive treatment of the Hindus by the Portuguese may have alienated the Brahmin keepers of the Kerala mathematics. The Brahmins were also possessive of this knowledge and only a privileged few were allowed access to its explications. Thus, as mentioned earlier, when Jesuit scientists like Rubino attempted to seek out this knowledge they were unsuccessful. Even if the Kerala mathematics reached Europe, attitudes of colonial supremacy in which Europeans saw themselves as 'civilizers' of the pagan colonies would have rejected the worth of the knowledge of 'barbarians'. That such an attitude existed is not difficult to accept given its later manifestations in the works of historians of science such as Sedillot (Almeida and Joseph 2004). If the conclusions of this research are confirmed in time *and* the reasons given here are valid, then this particular passage in the history of mathematics will be doubly noteworthy.

Case against Transmission: Ugo Baldini (2009)

Baldini (2009) has provided some of the more cogent arguments against the transmission hypothesis relating to the Jesuits. The arguments are based on showing a lack of motivation, opportunity and evidence (both circumstantial and documentary) for such a 'hidden migration' of ideas. He not only attempts to explain why, prior to 1650, there has been no known cases of scientific exchanges between Europe and India, but also examines where to focus future research on finding documentary evidence for these exchanges (if they had occurred). However, the reasons given for the lack of documentary evidence would make it unlikely that one would find such evidence in the future.

Baldini begins his paper by expressing surprise at the absence of Jesuits as scientists in India, particularly since there was a greater Jesuit presence on the subcontinent than in either China or Japan. By the end of the sixteenth century, the Jesuits had set up an extensive school and college system in both Goa and Cochin. And while there was the awareness among the Jesuit hierarchy in Rome that 'mathematics' (in the broad sense of the word) may be an appropriate key to open the 'locked doors', such an awareness was confined to mainly China. Baldini suspects that this was more due to 'some deep differences (true or false) perceived as existing between China's and India's socio-cultural reality' rather than the lack of appreciation of Indian mathematics or the absence of 'a significant figure like Matteo Ricci'.[19]

In testing transmission hypothesis, Baldini suggests that the following propositions need to be examined critically:

1. test the feasibility 'that one or more Jesuits *could* get (or, in fact, got) an *inner* knowledge of some of the more difficult parts of Kerala's mathematical works, little known even among the local students';
2. document (or to suppose) that this knowledge was transmitted to Europe;
3. 'document (not to suppose) that it reached some of the founders of the new mathematics. Obviously again, to document (3) makes (1) and (2) real, even if no proof of them is found; to document (2) makes (1) real, even if no proof of it is found, but it makes (3) only more plausible; to document (1) only makes (2) and (3) more plausible. So only a documentary proof for (3) may transform the transmission hypothesis into reality.'

Baldini proceeds to focus on two different points: the general possibility for some Jesuits to become acquainted (and to understand fully) some advanced Kerala mathematical texts; and the kind of Jesuit records which could be the vehicle for such a transmission. From a study of about 100–200 Jesuits who arrived in India from 1578 (the year that Matteo Ricci arrived in India) to 1640 as the upper limit beyond which transmission was unlikely to have influenced mathematical advances in Europe, he identifies a list of those who had the necessary mathematical skills (provided to them in Europe before leaving) to understand the content of, what was for the period highly sophisticated, Indian mathematics. The presupposition was that one or more of about 20 Jesuits who had had some mathematical training to be able to comprehend Kerala's advanced mathematics and had also spent some time in India could have been the carriers of mathematical knowledge which they then proceeded to transmit to Europe mainly through their contacts in the mathematics department of the Collegio Romano. Ideally, such a person would also have the necessary linguistic skills (in both Sanskrit and Malayalam) to 'extract the mathematical substance out of passages given in poetical/allegorical and obscure (for an European) modes of expression and (adopting) non-canonical (for an European) demonstration procedures'. Or failing that, a lower prerequisite was to 'imagine a person (that) could have the texts translated by another (European or Indian)' or be informed by local scholars of certain 'conceptions, methods and results during conversations with local students' (Baldini 2009: 284). Baldini (2009) argues that the scenario painted would be improbable:

> As evident, the ideal situation implies a number of circumstances which could hardly occur together: a Hindu—presumably a Brahmin—should have allowed a foreign person, who was also his rival in religion, to get in touch with precious and reserved texts; the other person should have been highly proficient in both Sanskrit and mathematics; he should also have been so patient and curious about those texts' content to overcome the barrier produced by unusual expressions and demonstration procedures. In front of this, it must be recalled that, as a matter of fact, no hint about Indian mathematical texts—perhaps some elementary ones excepted—is found in Jesuit sources before 1650 at least, and no such qualified person seems to have existed among India missionaries in those years; in particular, only De Nobili and, perhaps, Fenicio (both far from being mathematicians) seem to have been proficient in Sanskrit, while Rubino—for instance—only mastered Telugu and Malayalam.

Indeed, Baldini argues that the standard of mathematics in Jesuit schools rarely went beyond the elementary level and it would have been very difficult for most Jesuits to appreciate the 'refined mathematics like that of the Keraḷa School'. We are left with only a small list of Jesuits who had the mathematical background to become possible agents of the transmissions. Such a list would include Matteo Ricci (1578–1582), Muzio Rocchi (1597–1601/2), Carlos Spinola (1599–1601), Antonio Rubino (1602–1638), Sabatino De Ursis (1602–1603), Johann Schreck (1608–1609), Giulio Aleni (1608–1610), Cristoforo Borri (1615–1616, 1623–1624), Johann Chrysostoms Gall (1629–1643) and Hendrick Uwens (1647–1657). The names underlined are those who are often referred to in any discussion of transmission. For example, as documented earlier in the chapter, Ricci tried to get information about the local calendar on behalf of his teacher Clavius. Rubino admitted failure during his seven-year stay to find a suitable local informant on Indian astronomy. In the case of Schreck, there is tantalising evidence that the observations made of the comets of 1618 by him and other Jesuits such as Kriwitzer, Bell and Rho during their journey to India was passed to

Jesuit mathematicians in Inglolstadt who in turn communicated it to Kepler and who published this information in his work *Observationes Cometarum anni 1618 factae in India Orientali a quibusdam Soc. Jesu mathematicis in Sinense regnum navigantibus*, in 1620.

Baldini's main conclusion:

> Thus, unless new evidence is found and some basically new circumstance is established, the only possible deduction seems to be that not only no information exists on a Jesuit mathematician having managed to study some advanced Indian text (not to say to transmit it, or its content, to Europe), but no serious clue appears of a scientific interchange not purely superficial and more than occasional. (Baldini 2009: 294)[20]

Transmission Without Documentary Evidence: Arun Bala (2009)

Arun Bala argues the absence of documentary evidence is often the case where skill-based knowledge can be transmitted without documentation and so a search for documentary evidence alone may be misguided. It is quite possible that knowledge developed by the Kerala School may have been passed on as rules of computation. And, unlike the 'Greek and Platonic conceptions of mathematics as discovering truths about the world (or the ideal world of forms), the Indians saw their mathematical rules as analogous to the rules of grammar discovered by Panini (i.e., rules that defined correct practices rather than true beliefs)'. It was 'this flexible and pragmatic search for better computational techniques that led Indian mathematicians to the discovery of the Indian number system,[21] negative numbers, infinite series representations of irrational numbers, logarithms, trigonometry and mathematical series representations of circular functions'. And it is precisely because of this Indian approach to mathematics as computation techniques that Bala (2009) thinks that it 'may be looking in the wrong direction when one seeks evidence for transmission in letters, documents, translated texts etc. communicated by Jesuit missionary scientists and scholars to their counterparts in Europe'. '(For) ... the Jesuits would not see such knowledge as worthy of communication *qua* being mathematical knowledge'. It would have appeared to them as having only utilitarian value—the kind of practical rule-of-thumb knowledge that would interest only cartographers, mariners and calendar makers. But this would not have stopped them from using their training in mathematical and the Indian languages to translate such Indian mathematical techniques for use by their own cartographers, navigators and calendar makers.'

In other words, what Bala is suggesting is 'that while there may have been no obstacle to the assimilation of the discoveries of the Kerala School as technical knowledge by European craftsmen, there would have been barriers to transmitting such knowledge to European scholars as a form of intellectual knowledge. It lay in the epistemological divide which separated European conceptions of mathematics from the Indian conceptions.' What is being suggested here is 'the possibility that the Indian mathematical discoveries may have reached Europe as a set of practical computing rules rather than a body of mathematical discoveries'. This would be quite consistent with the fact that the 'mathematical discoveries of the Kerala School were practical empirical knowledge of techniques needed to make astronomical calculations—calculations used to determine religious holy days, design calendars that could predict likely times of arrival of the monsoons and so decide probable times for sowing and harvesting crops, and make charts for use by navigators and mariners'.

Now if such a direct transmission of such know-how from their Indian counterparts to European cartographers, navigators and calendar makers was allowed, there would be no need for any Jesuit mediation. 'The Indian computing techniques would disseminate widely simply because they were better than those available to the Europeans. If the most significant contribution of the Kerala School was moving from the finite procedures of ancient mathematics and treating their limit passage to infinity, then for practical computation purposes it is necessary to move from the infinite series representations to a finite series approximation. The transmitted finite series could have been used by European mathematicians to reconstruct the original infinite series discovered by Indian mathematicians.'

In the process of rediscovering the rule European mathematicians would not be aware of the contributions of their Indian counterparts to their own accomplishments. The process of rediscovery would have precluded such recognition. First, they did not receive the rule for the infinite series directly but reconstructed it from a finite series used as a practical rule-of-thumb by European craftsmen. Second, the finite series they reconstructed had itself been modified to fit Western trigonometric functions which were slightly different from the Indian equivalents. Finally, they would have been prepared to acknowledge discovery of an infinite series expansion as mathematically significant only after demonstrative proof for it [even if there was controversy about the adequacy of the proof]. Hence, on all of these grounds, the Indian influence on them would have been invisible to European mathematicians.

Conclusion

The possibility of independent European discovery of some of the Kerala mathematics is always there, although the choice of that as a default solution by most historians is debatable. Indeed without documentary evidence in the form of direct translations of Indian texts, or direct acknowledgement of Indian sources for these discoveries, the legal criterion carries a crucial weakness when it is extended to apply to the context of establishing scientific and mathematical discoveries. Since these are discoveries concerning an objective world, or the best techniques for solving problems, it is always possible for different people to make the same discoveries. Hence the similarity in the discoveries made need not necessarily imply influence—it may simply be the outcome of the independent discoveries of the best description of the world or best way of solving a problem. Since what works does not depend merely on our arbitrary choices, but is dictated by objective constraints, we might even expect this to occur. For example, many cultures have independently discovered the right-angle theorem attributed to Pythagoras in the West, but this is not the outcome of diffusion of knowledge but objective states of affairs. They were independent discoveries that yielded the same knowledge needed for constructing buildings in all cultures.

The central hypothesis of Bala's work is that if there was transmission of knowledge of infinite series to Europe, it was done indirectly through practical uses, with a truncated version being passed on from local craftsmen to their foreign counterparts (such as navigators) and then being reconstructed in Europe by the mathematically knowledgeable without being aware of its provenance. A question may be raised as to how it is was possible to transmit the knowledge of higher mathematics like the infinite series, through computations and calculations contained in navigation charts and similar aids, and if so what would be the precise nature of the calculation with series that could be transmitted. Within Europe of that time, there had already been established a tradition of tables that received a considerable boost with the establishment of printing press that had assured the availability of these tables to a wider population. Only a closer look at ships records and other practical manuals would help to resolve the validity of this hypothesis.

Another question that may be raised is how necessary was it to obtain increasingly accurate values for sine, cosine and of the circumference from a practical point of view. Opinions are divided as to whether accuracy (up to 10 or more decimal places) was required for navigational purposes. Of course, a 'delight in accurate calculation' may have driven the Kerala mathematicians to attempt increasingly accurate approximations.

The whole issue of 'internal' transmissions as a prelude to crossing the boundaries is of central importance. It could be argued that there is a need to look at the background of people involved, not just members of the Kerala School or their patrons from the royal courts, but also members of '*aharga*' who were more likely to be in closer contact with practitioners on matters of a commercial or navigational nature. Evidence need to be sought of the schools set up by the Portuguese and whether the medium of instruction was Malayalam, and also whether there were significant intellectual contacts between the local people and the Jesuits. This was not so in the case of Goa where the schools were conducted in only Portuguese. Evidence of Jesuit mastery of local languages, of Jesuit translations into Malayalam of texts of scriptures or training of local people to teach in schools would also be pertinent information.

It is worth remembering that if the hypothesis of transmission through navigational charts and cartographical calculations is to be sustained, one should take account of the fact that the Portuguese navigators who reached Kerala at that particular time possessed less expertise in these aspects than the North-Western Europeans who excelled in map-making and so on. This raises a problem in historical logistics. However, in the case of the Jesuits, they were recruited from all over Europe, with a disproportionate number originating from Central and North-West Europe.

There is a further issue in emphasising non-textual transmission which involves the question of the distinction between practical mathematics and higher mathematics since, by its very nature, oral transmission involved a lower level of mathematics. But then there is a deeper question on whether such a distinction is valid and in any case what was the level of mathematics required for an exercise involved in getting greater accuracy of tables using partial corrections and other approximation procedures.

Finally, it should be remembered that the question of transmission hinged to a degree on the nature of socio-cultural set-up of Kerala at that particular time. The general assumption is that the Brahminical class contributed to the study of higher mathematics. Was it possible that the other classes of society apart from the Brahmins, like the Ganaka, may have passed on the knowledge in vernacular language to the Portuguese? The extent of the symbiotic relationship between the Nambuthiri Brahmin and the Ganaka (found among the local practitioners) in the practice of astrology and medicine would support this point.

Notes

1. Indeed the importance of such techniques can be appreciated by the fact that many European governments instituted huge prizes for the discovery of accurate techniques of navigation—including the Spanish government in 1567 and 1598 and the Dutch in 1632—and leading scientists, including Galileo, competed for them.
2. The subject has also been examined in greater detail in earlier studies, notably see Almeida and Joseph (2004) and the Aryabhata Group (2002).
3. There are historians of science such as Joseph Needham (1969: 83) who argue that since 'details of any transmission are difficult to observe', it is reasonable to base claims for transmission on the observed fact that an idea or practice occurred in one place before it appeared in another as long as there was a corridor of communication between the two places. Needham also adds that it rests with those who claimed independent discovery to demonstrate their case—a practice that is hardly followed in scholarly circles.
4. The Jesuits are generally perceived as the mediators of Western science, especially in China. What is indicated here is another role: as intelligence-gatherers to fill some critical gaps that had appeared in Renaissance Europe, namely navigation and calendar construction (see The Aryabhata Group (2002): 35–38 for more details). They will not be examined here.
5. See, for example, Iannaccone (1998) and Baldini (1992). Baldini states 'It can be recalled that many the best Jesuit students of Clavius and Geienberger (beginning

with Ricci and continuing with Spinola, Aleni, Rubino, Ursis, Schreck, and Rho) became missionaries in Oriental Indies. This made them protagonists of an interchange between the European tradition and those Indian and Chinese, particularly in mathematics and astronomy, which was a phenomenon of great historical meaning' (Baldini 1992: 70; [my translation]).

6. See, for example, Wicki (1948-:, vol. IV: 293; vol. VIII: 458).
7. In the folio Goa 58, Jesuit library, ARSI, Rome.
8. For details see, The Aryabhata Group (2002: 33–48).
9. The position of Clavius in the history of science is assured by his role in the Gregorian calendar reform and his many publications. Clavius also maintained contacts with scientists and mathematicians outside Rome. Furthermore, Clavius was influenced by Arab mathematicians—see E. Knobloch, *Christoph Clavius (1538-1612) and His Knowledge of Arabic Sources*, http://www.ethnomath.org/resources/knobloch.pdf
10. According to M. Gorman (2003), Grienberger was '...a revisor of mathematical works written by Jesuits and in his strategies of engagement in epistolary relationships with natural philosophers and mathematicians outside the Jesuit order'(pp. 1–120).
11. See The Aryabhata Group (2002).
12. A letter from Mersenne to Buxtorf. Mersenne mentions the knowledge of Brahmins and 'Indicos'.
13. A letter from Mersenne to Rivet.
14. For details see, Almeida and Joseph (2004).
15. See E. Knobloch, *Christoph Clavius (1538-1612) and His Knowledge of Arabic Sources*, http://www.ethnomath.org/resources/knobloch.pdf
16. See, for example, Berggren (1986) and Rashed (1994: 143–146).
17. From the very beginning, the founder of the Jesuit Order, Ignatius Loyala, saw that the missionary labours of the Jesuits should be primarily directed among the 'pagans' of India, Japan, China, Canada, Central and South America. Only with time did their attention turn to other Christian countries. As the object of the society was the propagation and strengthening of the Catholic faith everywhere, the Jesuits naturally endeavoured to counteract the spread of Protestantism. They became the main instruments of the Counter-Reformation; the re-conquest of southern and western Germany and Austria for the Church, and the preservation of the Catholic faith in France and other countries were due chiefly to their exertions.
18. Polanco, an assistant to Ignatius, laid down that three things had to be considered regarding any letters sent to Rome. First, what was to be written; second, how it was to be written; third, with what diligence it was to be written and despatched. Polanco also gave detailed directions about the distribution of the letter and who had access to them. For further details see Correia-Alfonso (1969: Chapters 2–4).
19. Baldini points out that apart from a few episodes—such as the conversations with the Mughal Akbar, or King Venkata II of Vijaynagara or some learned Brahmins (in case of Nobili later)—there was no record of a Jesuit scientific enterprise aimed at impressing the Indian elite.
20. Baldini also hypothesised that, during the navigation to India or while in Goa or Cochin, some of the specialists taught some advanced mathematics courses to their fellow brothers, so that some of the latter could become proficient enough to cope

with algebra, progressions and series. There is proof that such courses were taught *occasionally* during navigation, in the Indian colleges and in Macao. However, the information found on those courses usually mentions basic astronomy and pure and applied geometry, not advanced computation.

21. This is a point that we discussed in the last chapter when considering the rise and spread of the Indian numeral system.

10

A Final Assessment

In the history of mathematics, the invention of calculus and 'the passage to infinity' are seen as major benchmarks in the creation of modern or advanced mathematics. The important question that this book addressed is the possibility of the beginnings of modern mathematics to have occurred outside Europe and these ideas and techniques in turn then transmitted to Europe. If such transmission proves to be significant, it would help to deconstruct the prevailing Eurocentric account of the development of modern mathematics.

Generally speaking, certain mathematical facts of a fundamental but elementary nature can be and are discovered independently by many cultures. In antiquity very often it *was a specific application* of a particular mathematical technique (perhaps conveyed in the form of just a recipe, *with no further conceptual justification*, for solving some problem of mensuration or geodesy) that might have been communicated from one culture to another.

In turn, the acquisition of this knowledge might have stimulated the mathematicians of the recipient culture to examine the theoretical basis of the recipe. Such an examination could lead not only to the independent rediscovery of the same mathematical fact by the recipient, but could sometimes lead to a related extension that might not have occurred to the former culture. In general, the subject of transmission directions and mechanisms is a complex one, fraught with possibilities for error, due to the paucity of the record and the difficulty of comprehending the texture of social life in a historical era from the written record.

There are two related questions as to how mathematical ideas originate and develop and how such ideas get transmitted from one culture-area to another. This book discussed both. There are widespread culture-centrist claims made regarding the direction of transmissions (often involving some form of retrospective privileging[1]). But the methodology underlying the testing of such claims and assessing the relevant evidence remains relatively undeveloped. This book attempted to do so by examining a distinct period of Kerala history—the period between fourteenth and seventeenth century AD.

The particular issue of Kerala mathematics raises some difficult methodological issues that needs to be addressed. If transmission is established, the question arises as to whether such a transmission took the form of movement of ideas or of practices or of both. After Needham's work on Chinese science, there has been a significant change in the attitude to Chinese civilisation and this in part arose from the recognition of the importance of the interaction between Chinese science and technology. A similar reappraisal in the Indian case is long overdue.

There are different dimensions to the process of transmission. Transmissions have taken place historically in India both within and outside as well during periods before and after the establishment of the British Raj. These need to be examined within the historical context of their occurrence.

Only one conduit of transmission and one set of data—namely the activities of Jesuits—are explored in this book. There is need to investigate other possible conduits within a wider canvas, including those practices in areas such as cartography, navigation, trade, activities of technicians and craftsmen.[2]

A whole number of issues are exposed with the opening up of the Pandora's Box of multiple conduits from Kerala. These include the process of 'underground' transmission through navigators or craftsmen, the importance and presence of the local *ganaka*s (astrologers) as mediators between the Brahmins and the populace, especially in the construction and use of *panchagam* (astrological calendar), the Arab traders as transmitters, the limitations of a view of top-down transmissions when there is always the possibility of knowledge at the bottom being transmitted above. Systems of knowledge should be emphasised not in terms of what happened to the educated but also how knowledge was used from other groups. Indeed, recent scholarship that has looked at scholar–craftsmen contacts within Europe in the context of the role of experimental philosophy in facilitating the birth of modern science could be extended to include the possibility of Europeans coming into contact with non-European ideas and practices, transmitted to European craftsmen by their non-European counterparts whom they came in contact with through their voyages of discovery in the early modern era.

Innovations take place both at level of practice as well as the level of theory. Indeed, practice [transmitted through craftsmen] could be as important as theory in the case of movement of ideas and technologies across cultures. Further, identification can structure theory, so that knowledge transmitted through craftsmen across cultures result in the recipient craftsmen developing theories on their own which in turn could inspire others more technically or theoretically qualified to construct their theories. Thus, for example, the transmission of a mathematical practice originating in Kerala through the navigators could in turn be provided their peculiar gloss by the European navigators, who in turn may inspire the European mathematicians eventually to develop their own mathematical explanations when none existed before, which bear little or no resemblance to the original product.

This book highlighted the importance of examining in detail the socio-economic context in which Kerala mathematics developed. The important issue is not who did it first, but the historical conditions in which this mathematics arose in a subcontinental context. The lines of communication within the Europe of the period were extensive while the geographical and size constraints in India tended to restrict communication.

There is also a further question on why in Kerala the period between the thirteenth and sixteenth centuries was so productive, resulting in a vast amount of literature, compared to other parts of South India, not only in mathematics and astronomy, but in practical arts (*arthasastra*), on spiritual (*bhakti*) traditions and on medicine (*ayurveda*) apart from literary compositions. Could this be explained by increase in state patronage? Did the development of Malayalam as a written language during this period have any bearing on the emergence of this 'golden period' in Kerala?

Studies of the Jesuits as scientific mediators/missionaries in India on the basis of Jesuit archives in India have been sparse, consisting of no more than 20 items. The actions of Marquess de Pomba, in 1759, of confiscating Jesuit possessions and destroying Jesuit records have not helped. More studies have been made regarding activities in China with the Jesuits there being in charge of the Astronomical Bureau and acting as scientific advisers for about two centuries. As far as India is concerned, what remains are mainly astronomical observations made during the eighteenth century and general descriptions of the country. None of the 'missionary' archives either in India or outside contain scientific information. However, there are papers in the Jesuit Archives at Kodaikanal to indicate the close relationship between members of the Court of Cochin and the Portuguese.

Mathematics is essentially a product of the culture in which it originated and it was not, until the advent of modern mathematics, a single, unified entity. Therefore, a historian of pre-modern mathematical texts, whether written in Akkadian, Arabic, Chinese, Egyptian, Greek, Persian, Sanskrit or any other language of a specific culture, must avoid the temptation to conceive of these as early attempts to express modern scientific ideas.

However, the individuality of the pre-modern mathematical traditions of the older non-Western societies is highlighted in this book from the examination of one of these traditions in the Indian context in some detail. In doing so, the author became aware fairly early on of the dangers of oversimplifying complex historical phenomena. Consider, for example, the characteristics of the Indian *shastra* of *jyotisa*.[3] This *shastra* was conventionally divided into three sub-groups: *ganita* (mathematical astronomy and mathematics itself), *samhita* (divination, including by means of celestial omens) and *hora* (astrology). A number of *jyotisi*s (that is, students of the *jyotisa*s) followed all three branches, a larger number just two (usually *samhita* and *hora*), and the largest number just one (*hora*).

The principal writings in *jyotisastra*, as in all Indian *shastras*, were normally in verse, though the numerous commentaries on them were almost always in prose. The verse form with its metrical demands, while it aided memorisation, led to greater obscurity of expression than prose composition would have entailed. Moreover, numbers had to be expressible in metrical forms (the two major systems used for numbers, the *Bhutasankhya* and the *Katapyadi*, were explained in an earlier chapter), and the consequent ambiguity of these expressions encouraged the natural inclination of Sanskrit *pandit*s (scholars) to test playfully their readers' acumen. It took some effort and practice to achieve accuracy in discerning the technical meanings of such texts. But in this opaque style the *jyotisi*s produced an abundant literature. Regrettably, only a relatively small number of these have been subjected to modern analysis, and virtually the whole ensemble is rapidly deteriorating and will soon be unavailable.

There is yet another problem. This arises from the tendency among some to claim that ideas that prevail at a particular point in time belong solely to their native cultures. To respond to this aspect of 'non-recognition', it has been suggested that one should view 'intercultural exchanges' as periods of revolutions. Like in periods of scientific revolutions, intercultural exchanges may involve a change of worldview, a change of meaning or even a change of the repertoire of questions. Intercultural exchanges cannot be conceived merely as an accumulation of past cultural experiences but should be regarded as new entities. Thus Tibetan astronomy (incorporating both Indian and Chinese influences), for example, cannot be recognised as merely an extension of Chinese or Indian astronomy, unless as a shorthand; it provides a useful illustration of how different traditions may join forces in the development of science.[4] Neither is the European Renaissance (with its Greek, Medieval and Islamic influences) a continuation of medieval culture. The fact that the propagators of a new cultural product belong predominantly to a specific ethnic group or that the culture thrives mainly in a certain country are not proofs of entitlement. The former will marginalise the effort of the minority who contributes to the new product while the latter is building on the fallacy of the former, that is, the country of origin supersedes the contribution of the minority. With globalisation, intercultural exchanges take place on a wider scale and the politics of recognition will escalate. Once we recognise that no one culture can claim sole ownership over intercultural products, we can provide an effective response to a common flaw in the discourse on monoculturalism and multiculturalism.

Intuitively, we associate cultures with certain ethnic groups or with countries of origin and it is the most natural way to speak of cultures. What is hinted as an important theme of this book is how problematic the ownership of cultures can be. 'Intercultural products' include, but need not necessarily be limited to, a new philosophy, a new form of literature or a new scientific temperament produced in the aftermath of an intercultural exchange. The claim to the entitlement of a 'cultural hybrid' is part of the politics of recognition.[5] And the demand and the need for recognition remains one of the propelling forces for contemporary politics. Ethnic groups and nations have been suppressed, injured and gone to war because of misrepresentation. It is a problem worth considering because of the extent and frequency of intercultural exchanges in the course of human history, and thus, a major source of conflict among various ethnic groups.

At the root of this conflict is the standard definition of culture as 'a common language, a shared history, a shared set of religious beliefs and moral values, and a shared geographical origin, all of which taken together define a sense of belonging to a specific group'. This essentialist model of culture (that is, there are unchanging essences in a particular culture) has been popularised by buzz-words encompassed in the title of Huntington's work, *The Clash of Civilizations.* But following on from Thomas Kuhn's insights, it would be misleading to view a culture as an accumulation of cultural achievements by a particular group. With each epoch, especially one heralded by an external culture, there is a radical departure from thinking of previous periods. There may be a period in history when cultures may be properly denoted as belonging to an ethnic group but with each intercultural exchange, these cultures are deeply transformed. Thus, our popular use of terms like 'Islamic culture', 'Western culture' and 'Chinese culture' should be highly qualified and only as shorthand for a complex reality.

The question of ownership of 'cultural hybrids' is not merely an issue in monoculturalism. The dominant culture takes the position that inventions and discoveries from other cultures are not significant until they are assimilated into the dominant culture. It has been said that though the Chinese invented paper, gunpowder, printing and the magnet, but it was the Europeans who truly utilised them. Similarly, while acknowledging the Islamic and the Indian scholars as pioneers of our numeral system, little attention has been given to how the discovery was made and much has focused on the extent to which mathematics has been applied to modern science. The gaps in the history of mathematics perpetuate the myth of a dominant culture. The essentialist model of culture is most prominent in monoculturalism as it only allows one cultural narrative. To portray a sense of unity, foreign elements are assimilated within the official history and if that fails, the alternative accounts are marginalised.

Attempts to correct historical myopia have been hindered by the assumption of universality. The transcendent nature of mathematics has been emphasised to the obliteration of the locality of its origin. I am not a relativist but to a certain extent, universality has meant Westernisation, much like how 'man' has been used to denote humanity but is in fact a disguise for the stereotypical male. This is how mathematics has been viewed currently, often as a creation of the Western world, rather than a creation of various civilisations. It implies that other mathematical and scientific traditions were in its primitive and unsystematic stages and history of mathematics and the sciences should properly begin in the modern era. With this paradigm encouraging mathematicians to view history from Western perspectives, research into mathematics from other traditions had been stymied until recent decades. It is hoped that this book provides an antidote to the expression of any overt manifestation of the politics of recognition where one mathematical tradition takes the credit for all significant advances in the subject.

Notes

1. 'Retrospective privileging' implies 'looking back' and granting a special place or privilege to a certain group or culture. The most powerful and sustained form of retrospective privileging has been the Eurocentric Vision, the centre-piece of which are the existence of a 'Greek Miracle' and the appearance of the 'Dark Ages'.
2. There were, of course, ideas-mediated transmission through Arab and Chinese intermediaries to Europeans which need to considered. Such studies are very much in their infancy.
3. *Jyoti* is a Sanskrit word meaning 'light', usually from a 'star'; so that *jyotisastra* means 'teaching about the stars'.
4. Tibetan astronomy is also valuable for the light it has thrown on the history of Indian astronomy before the appearance of the *Siddhantas* in the fifth century AD. And it provides one of the few examples of Buddhism influencing the growth of Indian astronomy, for the bulk of evidence for the spread of Indian astronomy to Tibet comes from texts from a period that saw the decline of Buddhism as a religion in India. For further details, see Joseph (1994b).
5. For an interesting discussion of the ownership of 'cultural hybrids', see Lam Chek Wai on www.inter-disciplinary.net/ci/interculturalism/IC1/lau%20paper.pdf. The discussion that follows is based on this interesting work.

Bibliography

Abeles, F.F. (1993) 'Charles Dodgson's Geometric Approach to Arctangent Relations for π', *Historia Mathematica*, 20(2): 151–59.

Almeida, D.F., J.K. John, and A. Zadorozhnyy (2000) 'Keralese Mathematics: Its Possible Transmission to Europe and the Consequential Educational Implications', *Journal of Natural Geometry*, 20(1): 77–104.

Almeida, D.F. and G.G. Joseph (2004) 'Eurocentrism in the History of Mathematics: The Case of the Kerala School', *Race and Class*, 45(4): 45–60.

——(2009a) 'A Report of the Investigation on the Possibility of the Transmission of the Medieval Kerala Mathematics to Europe', in G.G. Joseph (ed.), *Kerala Mathematics: Its History and Possible Transmission to Europe*, pp. 257–76. Delhi: B. R Publishing Corporation.

——(2009b) 'Kerala Mathematics and its Possible Transmission to Europe', in P. Ernest, B. Greer and B. Sriraman, *Critical Issues in Mathematics Education*. Charlotte, NC: Information Age Publishing.

Al-Andalusi (c. 1068) *Science in the Medieval World*, translated by S.I. Salem and A. Kumar. Austin: University of Texas Press, 1991.

Apte, V.G. (ed.) (1937) *Lilavati of Bhaskara II*. Anandasrama Sanskrit Series, 107. Pune: Anandasrama Press.

The Aryabhata Group (2002) 'Transmission of the Calculus from Kerala to Europe', proceedings of the International Seminar and Colloquium on 1500 Years of Aryabhateeyam, Kerala Sastra Sahitya Parishad, Kochi, pp. 33–48.

Arundhati, P. (1994) *Royal Life in Manasollasa*. Delhi: Sundeep Prakashan.

Bag, AK. (1976) 'Madhava's sine and cosine sereis', *Indian Journal of History of Science*, 11(1): 54–57.

Bag, A.K. and S.R. Sharma (eds). (2003) *The Concept of Sunya*. Delhi: Aryan Books.

Bala, A. (2009) 'Establishing Transmissions: Some Methodological Issues', in G.G. Joseph (ed.), *Kerala Mathematics: Its History and Possible Transmission to Europe*, pp. 155–180. Delhi: B. R Publishing Corporation.

Balachandra, Rao S. (1994) *Indian Mathematics and Astronomy*. Bangalore: Jnana Deep Publications.

Baldini, U. (1992) *Studi su filosofia e scienza dei gesuiti in Italia 1540–1632* (Studies on Science and Philosophy of hte Jesuits in Italy, 1540–1632). Rome: Bulzoni Editore.

Baldini, U. (2009) 'The Jesuit Mathematicians in India (1578–1650) as Possible Intermediaries between European and Indian Mathematical Traditions', in G.G. Joseph (ed.), *Kerala Mathematics: Its History and Possible Transmission to Europe*, pp. 277–306. Delhi: B.R Publishing.

Baldini, U. and P.D. Napolitani. (1992) *Corrispondenza / Christoph Clavius; edizione critica a cura*. Pisa: Università di Pisa, Dipartimento di matematica.

Baron, M.E. (1987) *The Origins of the Infinitesimal Calculus*. Oxford: Pergamom.

Benedict, S.R. (1914) 'A Comparative Study of the Early Treatises Introducing into Europe the Hindu Art of Reckoning', Ph.D. Thesis, University of Michigan, Rumford Press.

Bentley, J. (1823) *A Historical View of the Hindu Astronomy*. Calcutta: Baptist Mission Press.

Berggren. J.L. (1986), *Episodes in the Mathematics of Medieval Islam*. New York: Springer-Verlag.

——(2007) 'Mathematics in Medieval Islam', in V. Katz (ed.), *The Mathematics of Egypt, Mesopotamia, China, India, and Islam, A Sourcebook*. Princeton: Princeton University Press.

Bernal, M. (1987) *Black Athena*. London: Free Association Books.

Bernard, H. (1973) *Matteo Ricci's Scientific Contribution to China*. Westport, CT: Hyperion Press, Al-Biruni (1030) *India*, translated by Qeyamuddin Ahmad, New Delhi: National Book Trust, 1999.

Boyer, C.B. (1959) *The History of the Calculus and Its Conceptual Development*. New York: Dover.

Bressoud, D. (2002) 'Was Calculus Invented in India?' *The College Mathematics Journal* (Mathematical Association of America), 33(1): 2–13.

Brezinski, C. (1980) *History of Continued Fractions and Pade Approximations*. London: Springer-Verlag.

Burgess, E. (1860) *The Surya Siddhanta: A Text-Book of Hindu Astronomy*, reprinted, 1997. New Delhi: Motilal Banarsidass Publishers Private Limited.

Burnett, C. (2002) 'Indian Numerals in the Mediterranean Basin in the Twelfth Century, with Special Reference to the "Eastern Forms"', in Y. Dold-Samplonius, J.W. Dauben, M. Folkerts and B. van Dalen (eds), *From China to Paris: 2000 Years' Transmission of Mathematical Ideas*, pp. 237–88. Stuttgart: Steiner.

Calinger, R. (1999) *A Contextual History of Mathematics to Euler*. New Jersey: Prentice Hall.

Cantor, M. (1892) *Vorlesungen über Geschichte der Mathematik*, 4 vols; vol. II, 1892, 2nd edition, 1900. Leipzig: Teubner.

Chandrasekharan, T. (1956) *Karanapaddhati of Putumana Somayaji*. Madras: Govt. Oriental Mss Library.

Colebrooke, H.T. (1873) *Miscellaneous Essays*, 2 vols. London: Allen and Company.

Correia-Alfonso, J. (1969) *Jesuit Letters and Indian History 1542–1773*. Bombay: Oxford University Press.

Cortesão, A. and L. de Albuquerque (1982) *Obras completas de D. Joao de Castro*, vol. IV. Coimbra: University of Coimbra.

Correspondance du P. Marin Mersenne. (1945-) *Correspondance du P. Marin Mersenne*, 18 volumes. Paris. Presses Universitaires de France.

de Curtin, P. (1984) *Cross Cultural Trade in World History*. London: Cambridge University Press.

Clark, W.E. (1930) *The Aryabhatiya of Aryabhatam*, translated into English with notes. Chicago: The University of Chicago Press.

Datta, B. (1927) 'Early History of the Arithmetic of Zero and Infinity in India', *Bulletin of the Calcutta Mathematical Society*, 18: 165–76.

——(1932) 'Elder Aryabhata's Role for the Solution of Indeterminate Equations of the First Degree', *Bulletin of the Calcutta Mathematical Society*, 24: 19–36.

Descartes, R. (1954) *The Geometry*, translated from French and Latin by David Eugene Smith and Marcia L. Latham. Dover: New York.

Deva Sastri, B. (ed.) (1989) *Siddhanta-Siromani*, Kashi Sanskrit Series, 72. Varanasi: Chauklamba Sanskrit Sanstan.

D'Elia, P. (1960) *Galileo in China: Relations through the Roman College between Galileo and the Jesuit Scientist-Missionaries (1610–1640)*, translated by R. Suter and M. Sciascia. Harvard University Press.

Dvivedi, S. (1933) *Garṇakataranginị*, revised edition. Benares: B.K. Shastri.

Edwards, C.A.H. (1979) *The Historical Development of the Calculus*. New York: Springer-Verlag.

Elfering, K. (1977) 'The Area of a Triangle and the Volume of a Pyramid as well as the Area of a Circle and the Surface of the Hemisphere in the Mathematics of Aryabhata I', *Indian Journal of the History of Science*, 12(2): 232–36.

Ernest, P. (2009) 'The Philosophy of Mathematics, Values and Keralese Mathematics', in G.G. Joseph *Kerala Mathematics: Its History and Possible Transmission to Europe*, pp. 181–202. Delhi: B.R. Publishing Corporation.

Eves, H. (1983) *An Introduction to the History of Mathematics*, 5th edition. Philadelphia: Saunders.

Fauvel, J. and J. Gray, (eds) (1987) *The History of Mathematics: A Reader*. London: Macmillan.

Fiegenbaum, L. (1986) 'Brook Taylor and the Method of Increments', *Archive for the History of Exact Sciences*, 34(1): 1–140.

Ganguly, P.C. (1928) 'The Source of the so-called Pellian Equations', *Bulletin of the Calcutta Mathematical Society*, 19: 151–76.

Gorman, M. (2003) 'Mathematics and Modesty in the Society of Jesus: The Problems of Christoph Grienberger (1564–1636)', in Mordechai Feingold (ed.), *The New Science and Jesuit Science: Seventeenth Century Perspectives*, Archimedes vol. 6. pp. 1–120. Kluwer: Dordrecht.

Govinda, Pillai P. (1955) *Malayala Bhasha Charitram*. Kottayam: National Book Stall.

Gupta, R.C. (1969) 'Second Order Interpolation in Indian Mathematics up to the Fifteenth Century', *Indian Journal of History of Science*, 4 (1&2): 86–98.

Hawksworth, A. (1860) *Day Dawn in Travancore*. Kottayam: CMS Press.

Hayashi, T. (1995) *The Bhakhshali Manuscript: An Ancient Mathematical Treatise*. Groningen: Egbert Forsten.

Hayashi, T., T. Kusuba and M. Yano, (1990) 'The Correction of the Madhava Series for the Circumference of a Circle', *Centaurus*, 33 (2&3): 149–74.

Heath, T.L. (1910) 'Diophantus' Method of Solution', in *Diophantus of Alexandria: A Study in the History of Greek Algebra*. Cambridge: Cambridge University Press.

Heeffer, A. (2007) *The Tacit Appropriation of Hindu Algebra in Renaissance Practical Arithmetic*. Belgium: Centre for Logic and Philosphy of Science, Ghent University.

Hooykaas, R. (1983) *Selected Studies in History of Science*, Coimbra: Por ordem da Universidade Coimbra.

Iannaccone, I. (1998) *Johann Schreck Terrentius*. Napoli: Instituto Universitario Orientale.

Ifrah, G. (2000) *The Universal History of Numbers: From Prehistory to the Invention of the Computer*. New York: Wiley.

Jami, C. (1988) 'Western Influence and Chinese Tradition in an Eighteenth Century Chinese Mathematical Work', *Historia Mathematica*, 15: 311–31.

Jesseph, D.M. (1999) *Squaring the Circle: The War between Hobbes and Wallis*. Chicago: University of Chicago Press.

Joseph, G.G. (1994a) 'Different Ways of Knowing: Contrasting Styles of Argument in Indian and Greek Mathematical Traditions', in P. Ernest (ed.), *Mathematics, Education and Philosophy: An International Perspective*, pp. 194–204. London: The Falmer Press.

——(1994b) 'Tibetan Astronomy and Mathematics', in I. Grattan-Guinness (ed.), *Companion Encyclopedia of the History and Philosophy 7 of the Mathematical Sciences*, pp. 131–36. London: Routledge.

——(1995) 'Cognitive Encounters in India during the Age of Imperialism', *Race and Class*, 36(3): 39–56.

——(1997) 'What is a Square Root? A Geometrical Representation in Different Mathematical Traditions', *Mathematics in Schools*, 23(3), 4–9, reprinted in C. Pritchard (ed.), *The Changing Shape of Geometry*. Cambridge: Cambridge University Press, 2003.

——(2000) *The Crest of the Peacock: Non-European Roots of Mathematics*. Princeton and London: Princeton University Press.

——(ed.) (2009) *Kerala Mathematics: Its History and Possible Transmission to Europe*. Delhi: B.R. Publishing Corporation.

Jushkevich, A.P. (1964) *Geschichte der Matematik im Mittelater*, German translation. Leipzig.

Katz, V.J. (1992) *A History of Mathematics: An Introduction*. New York: Harper Collins.

Kaye, G.R. (1908) 'Notes on Indian Mathematics', *Journal of the Asiatic Society of Bengal*, 4: 111–41.

Keller, A. *Expounding the Mathematical Seed: A Translation of Bhaskara I on the Mathematical Chapter of Aryabhatiya*, Vol 1. Basel: Birkauser Verlag.

Kline, M. (1972) *Mathematical Thought from Ancient to Modern Times*. New York: Oxford University Press.

Kunjunni, Raja K. (1963) 'Astronomy and Mathematics in Kerala', *Adyar Library Bulletin*, 27: 118–67.

Kuppanna Sastry, T.S. (1985) *Vedanga Jyotisa of Lagadha in its RK and Yajur Recensions*. Delhi: Indian National Science Academy.

Lam Lay Yong. (1986) 'The Conceptual Origins of Our Numeral System and the Symbolic Form of Algebra', *Archive for History of Exact Sciences*, 36: 183–99.

——(1987) 'Linkages: Exploring the Similarities between the Chinese Rod Numeral System and Our Numeral System', *Archive for History of Exact Sciences*, 37: 365–92.

Logan, W. (1906) *Malabar*, 2 vols. Trivandrum: Charithram Publications.

Li Yan and Du Shiran (1987) *Chinese Mathematics: A Concise History*. Oxford: Clarendon Press.

Madhavan, S. (1991) 'Origins of Katapayadi System of Numerals', *Sri Rama Varma Samskrta Granthavali Journal*, 18(2): 35–48.

Majumudar, N.K. (1912) 'Aryabhata's Rule in Relation to Indeterminate Equations of the First Degree', *Bulletin of the Calcutta Mathematical Society*, 3: 11–12.

Mallayya, V.M. (2002) 'Geometrical Approach to Arithmetic Progressions from Nilakantha's Aryabhatiyabhasya and Sankara's Kriyakramakari', proceedings of the International Seminar and Colloquium on 1500 Years of Aryabhateeyam, Kerala Sastra Sahitya Parishad, Kochi, pp. 143–47.

——(2004) 'An Interesting Algorithm for Computation of Sine Tables from the *Golasara* of Nilakaṇṭha', *Ganita Bharati*, 26: 40–55.

——(2009) *Trignometric Sines and Sine Tables in India: A Survey*. A Preprint, AHRB Project, University of Manchester.

Mallayya, V.M. and G.G. Joseph (2009a) 'Indian Mathematical Tradition: The Kerala Dimension', in G.G. Joseph (ed.), *Medieval Kerala Mathematics: History and Its Possible Transmission to Europe*, pp. 35–58. Delhi: B.R. Publishing Corporation.

——(2009b) 'Kerala Mathematics: Motivation, Rationale and Method', in G.G. Joseph (ed.), *Medieval Kerala Mathematics: History and Its Possible Transmission to Europe*, pp. 77–112. Delhi: B.R. Publishing Corporation.

Marar, K.M. and C.T. Rajagopal (1944) 'On the Hindu Quadrature of the Circle', *Journal of the Bombay Branch of the Royal Asiatic Society*, (NS) 20: 65–82.

Menon, P.R. (1931) *Laghu Ramayana*, Tuncattu Granthavali, No. 3, Chittoor.

——(1952–1953) 'Tuncattu Ezhuttacchan', *Tuncattu Ezhuttacchan* Monthly, III(3): 127–35.

Menon, S. (1979) *Social and Cultural History of Kerala*. New Delhi: Sterling Publisher.

Narayana, Panickar (1951) *Kerala Bhasha Sahitya Charitram*, 7 vols. Trivandrum: Vidyavilasini Book Depot.

Narayanan, M.G.S. and Kesavan Veluthat (1983), 'A History of the Nambudiri Community in Kerala', in F. Staal (ed.), *Agni*, vol. II. Berkeley: Asian Humanities Press.

Narayanan, Namputiri V. (1975) *Karaṇamrta* of Citrabhanu, Trivandrum Sanskrit series No. 240, Trivandrum: University of Kerala.

Needham, J. (1959) *Science and Civilization in China*. Vol 3. Cambridge: Cambridge University Press.

——(1969) *The Grand Titration: Science and Society East and West*. London: Allen and Unwin.

Neugebauer, O. (1962) *The Exact Sciences in Antiquity*. New York: Harper.

O'Leary, D.L. (1948) *How Greek Science Passed to the Arabs*. London: Routledge and Keegan Paul.

Peacock, G. (1849) 'Arithmetic—including a History of the Science', in *Encylcopedia Metropolitana or Universal Dictionary of Knowledge*; Part 6, First Division. London: J J Griffin and Co.

Pillai, Suranad Kunjan (1957) *Aryabhaṭīya bhaṣya* of Nīlakantha, *Golapada*, Trivandrum Sanskrit series No. 185. Trivandrum: University of Kerala.

Plofker, K. (2001) 'The "Error" in the Indian "Taylor Series Approximation" to the Sine', *Historia Mathematica*, 28: 283–95.

Prag, A. (1939) 'On James Gregory's Geometriae Pars Universalis', in H.W. Turnball (ed.) *James Gregory Tercentenary Memorial Volume. Containing His Correspondence with John Collins and His Hitherto Unpublished Mathematical Manuscripts, Together with Addresses and Essays Communicated to the Royal Society of Edinburgh*. London: G. Bell & Sons.

Raghava Varier, M.R. (2002) 'Swarupam: An Introductory Essay', in Champakalakshmi R et al. (eds), *State and Society in Pre-modern South India*, pp. 120–30. Trissur: Cosmobooks.

——(2009) 'Uncatalogued Malayalam Manuscripts in Europe: A Report of Work Carried out in Two Libraries in Rome, in G.G. Joseph, (ed.), *Kerala Mathematics: Its History and Possible Transmission to Europe*, pp. 77–112. Delhi: B.R. Publishing Corporation.

Raja, K.K. (1963) 'Astronomy and Mathematics in Kerala', *Adyar Library Bulletin*, 27: 118–67.

Rajagopal, C.T. and Marar, M. (1944), 'On the Hindu Quadrature of the Circle', *Journal of the Royal Asiatic Society* (Bombay branch), 20: 65–82.

Rajagopal, C.T. and A. Venkataraman (1949) 'The Sine and Cosine Power Series in Hindu Mathematics', *Journal of the Royal Asiatic Society of Bengal*, 15: 1–13.

Rajagopal, C.T. and T.V.V. Aiyar (1951) 'On the Hindu Proof of Gregory's Series', *Scripta Mathematica*, 17: 65–74.

Rajagopal, C.T. and T.V. Vedamurthi (1952) 'On the Hindu Proof of Gregory's Series', *Scripta Mathematica*, 18: 65–74.

Rajagopal, C.T. and M.S. Rangachari (1978) 'On an Untapped Source of Medieval Keralese Mathematics', *Archive for the History of Exact Sciences*, 18(2): 89–102.

——(1986) 'On Medieval Keralese Mathematics', *Archive for the History of Exact Sciences*, 35(2): 91–99.

Ramasubramaniam, K. (2002) 'Aryabhateeyam in the Light of *Aryabhateeyam-bhashya* by Nilakantha Somayaji', proceedings of the International Seminar and Colloquium on 1500 Years of Aryabhateeyam, Kerala Sastra Sahitya Parishad, Kochi, pp. 115–22.

Ramasubramanian, K., M.D. Srinivas and M.S. Sriram (1994) 'Heliocentric Model of Planetary Motion in the Kerala School of Indian Astronomy', *Current Science*, 66: 784–90.

Ramavarma Maru Thampuran, Akhilesvara Aiyer (1952) *Yuktibhaṣa*, Part I. Trichur: Mangalodayam.

Rashed, R. (1994) 'Indian Mathematics in Arabic', in Ch. Sasaki, J.W. Dauben, M. Sugiura (eds), *The Intersection of History and Mathematics*. pp. 143–48. Basel, Boston and Berlin: Birkhaüser Verlag.

Rodet, L. (1879) 'Lecon de calcul d'Aryabhata', *Journal Asiatique*, Septieme Serie, 13: 393–434.

Saliba, G. (1994) *A History of Arabic Astronomy*. New York: New York University Press.

Sambasiva, Sastry K. (1930) *Āryabhaṭīya* with *bhāṣya* of Nīlakaṇṭha, Part I, Trivandrum Sanskrit series No. 101, Trivandrum.

——(1931) *Aryabhaṭiya* with *bhaṣya* of Nilakaṇṭha, Part II, Trivandrum Sanskrit series No. 110, Trivandrum.

Sarasvati Amma, T.A. (1963) 'Development of Mathematical Series in India', *Bulletin of the National Institute of Sciences (India)*, 21: 320–43.

——(1979) *Geometry in Ancient and Medieval India*. Delhi: Indological Publ.

Sarma, K.V. (1956) *Venvaroha* of Madhava with commentary by Acyuta Pisarati. Sanskrit College Committee, Tripunithura, Vishveshvaranand Institute.

——(1957) *Goladipika* of Paramesvara. Madras: Adayar Library.

——(1958) 'Jyesthadeva and his Identification as the Author of *Yuktibhasa*', *Adyar Library Bulletin*, 22: 35–40.

——(1963) *Drgganita* of Paramesvara. Hoshiarpur: Vishveshavaranand Institute.

——(1972a) 'Some Direct Lines of Astronomical Tradition in Kerala', P.C. Shastry Felicitation Volume. Delhi: Chaturveda.

——(1972b) *A History of Kerala School of Hindu Astronomy*. Hoshiarpur: Vishveshavaranand Institute.

——(ed.) (1972c) *Lilavati of Bhaskaracarya with Kriyakramakari of Sankara and Narayana*. Hoshiarpur: Vishveshavaranand Institute.

——(1974) *Sphutanirnaya Tantra* of Acyuta with Auto commentary. Hoshiarpur: Vishveshavaranand Institute.

——(1975) *Lilavati of Bhaskaracarya with the Kriyakramakari of Sankara and Narayaṇa*. Hoshiarpur: Vishveshavaranand Institute.

——(1977a) *Rasigolasphutaniti* of Acyuta. Hoshiarpur: Vishveshavaranand Institute.

——(1977b) *Tantrasangaha* of Nilakantha Somayaji with the *Yuktidipika* and *Laghuvivṛti* of Śȧnkara. Hoshiarpur: Vishveshavaranand Institute.

——(1985) *Vedanga Jyotisa & Lagadha* (in its R K and Yajus Recensious) (with the translational notes of Professor T.S. Kuppana Sastry). Delhi: Indian National Science Academy.

——(2001) *Sadratnamala* of Sankaravarman. Delhi: Indian National Science Academy.

——(2002) *Science Texts in Sanskrit in the Manuscripts Repositories of Kerala and Tamilnadu*. Delhi: Rashtriya Sanskrit Sansthan.

——(2008) *GanitaYuktiBhasa of Jyesthadeva*, 2 vols. New Delhi: Hindustgan Book Agency.

Sarma, K.V. and T.S. Kuppanna Sastry (1964) *Vakyakaraṇa* (*Vakyapancaddhyayi*). Madras: Kuppuswam Research Institute.

Sarma, K.V. and S. Hariharan. (1991) '*Yuktibhasa* of Jyesthadeva', *Indian Journal of History of Science*, 26: 186–207.

Sarma, S.R. (2002) 'Rule of Three and Its Variations in India' in Y. Dold-Samplonius J.W. Dauben, M. Folerts and B. van Dalen (eds), *From China to Paris: 2000 Years of Transmission of Mathematical Ideas*, pp. 133–56. Stuttgart: Steiner Verlag.

——(2003) 'Sunya in Pingala's Chandahsutra', in A.K. Bag and S.R. Sharma (eds), *The Concept of Sunya*, pp. 126–36. Delhi: Aryan Books.

——(2009) 'Early Transmissions of Indian Mathematics', in G.G. Joseph (ed.) *Kerala Mathematics: Its History and Possible Transmission to Europe*, pp. 35–58. Delhi: B.R. Publishing Corporation.

Scott, J.F. (1981 [1938]) *The Mathematical Work of John Wallis*. London: Taylor and Francis.

Sedillot, L.A. (1873) 'The Great Autumnal Execution', in the *Bulletin of the Bibliography and History of Mathematical & Physical Sciences*, published by B. Boncompagni, member of Pontific Academy, reprinted in *Sources of Science*, No. 10 (1964), New York and London.

——(1845–1849) *Sciences Mathematique chez les Grecs et les Orienteaux*, vol. II. Paris : Libraire de Firmin Didot Freres.

Sengupta, P.C. (1927) 'The Aryabhatiyam, Translated into English', *Journal of the Department of Letters*, 16: 1–56.

Shukla, K.S. (ed.) (1957) *The Suryasiddhanta with the Commentary of Paramesvara*. Lucknow: Department of Mathematics and Astronomy.

——(1960) *Mahabhaskariya*, edited and translated into English with explanatory and critical notes, and comments, etc. Lucknow: Department of Mathematics and Astronomy, Lucknow University.

——(1976) *Aryabhatiya of Aryabhata, with the Commentary of Bhaskara I and Somesvara*. Delhi: Indian National Science Academy.

Shukla, K.S. and Sarma, K.V. (1976) *Aryabhatiya of Aryabhata*. New Delhi: Indian National Science Academy.

Singh, A.N. (1936) 'On the Use of Series in Hindu Mathematics', *Osiris*, 1: 606–28.

——(1939) 'Hindu Trigonometry', *Proceedings of the Benares Mathematical Society*, 1: 77–92.

Smith, D.E. (1923/5) *History of Mathematics*, 2 vols. Boston, MA: Ginn & Co.)(Reprinted by Dover, New York, 1958.

Subbarayappa, B.V. and K.V. Sarma (1985) *Indian Astronomy—A Source-Book*. Bombay: Nehru Centre.

Sulloway, F.J. (1996), *Born to Rebel*. London: Little, Brown & Co.

Ulloor Paramesvara Iyer S. (1953–57) [1953] *Kerala Sahitya Caritram*, 3rd edition, 5 vols Trivandrum: Kerala University series 30, University of Kerala.

Unithiri, N.V.P. (2003) 'Astronomy and Mathematics in Medieval Kerala with Special Reference to the Nila Valley', in N.V.P. Unithiri (ed.), *Indian Scientific Traditions*. Calicut: University of Calicut.

Unithiri, N.V.P. and U.K. Anandhavardhan (2002) 'A Note on Nilakantha's Bhaysa on Aryabhitya', proceedings of the International Seminar and Colllogquium on 1500 Years of Aryabhateeyam, Kerala Sastra Sahitya Parishad, Kochi.

Vijalekshmy, M. and G.G. Joseph (2009) 'An Intellectual Background of Medieval Kerala with Special Reference to Mathematics and Astronomy', in G.G. Joseph (ed.), *Medieval Kerala Mathematics: History and Its Possible Transmission to Europe*, pp. 59–73. Delhi: B.R. Publishing Corporation.

van Der Waerden, B.L. (1976) 'Pell's equation in Greek and Hindu mathematics', *Russian Math Surveys*, 31: 210–25.

——(1983) *Geometry and Algebra in Ancient Civilizations*. Berlin: Springer-Verlag.

Wallace, W. (1984) *Galileo and His Sources: The Heritage of the Collegio Romano in Galileo's Science*. Princeton: Princeton University Press.

Warren, J. (1825) *Kala Sankalita: A Collection of Memoirs on the Various Modes According to which the Nations of the Southern Parts of India Divide Time*. Madras: College Press.

Whish, C.M. (1835) 'On the Hindu Quadrature of the Circle and the Infinite Series of the Proportion of the Circumference to the Diameter Exhibited in the Four Sastras, the *Tantrasangraham, Yukti-Bhasa, Carana Padhati*, and *Sadratnamala*', *Transactions of the Royal Asiatic Society of Great Britain and Ireland*, III(iii): 309–23.

Wicki, Josef (1948) *Documenta Indica*, 16 volumes. Rome: Monumenta Historica Societate Iesu.

Index

About the Author

George Gheverghese Joseph was born in Kerala, India. His family moved to Mombasa in Kenya where he did his schooling. He studied at the University of Leicester, the United Kingdom, and then worked for six years in Kenya before pursuing his postgraduate studies at Manchester, the United Kingdom. He has travelled widely, holding university appointments and giving lectures at various universities around the world. He has appeared on radio and television programmes in India, the United States, Australia, South Africa and New Zealand, as well as the United Kingdom. In January 2000, he helped to organise an International Seminar and Colloquium to commemorate the 1500th year of Aryabhata's famous text, *Aryabhateeyam*; the seminar was held in Thiruvanthapuram, Kerala. In December 2005, he organised an International Workshop at Kovalam which was the culmination of a AHRB-funded Research Project on 'Medieval Kerala Mathematics: The Possibility of Its Transmission to Europe'. In 2008, he gave talks at Loyola University, Chicago, USA and the Mathematical Association of America (MAA), reporting on the findings of the AHRB project. He holds joint appointments at the University of Manchester and at the University of Toronto, Canada.

He authored the bestseller *The Crest of the Peacock: Non-European Roots of Mathematics* (1991). His other works include *George Joseph: The Life and Times of a Kerala Christian Nationalist* (2003), *Multicultural Mathematics* (1993) and *Women at Work: The British Experience* (1983).